PARALLEL ARCHITECTURES AND PARALLEL ALGORITHMS FOR INTEGRATED VISION SYSTEMS

PARALLEL ARCHITECTURES AND PARALLEL ALGORITHMS FOR INTEGRATED VISION SYSTEMS

by

Alok N. Choudhary

Syracuse University

and

Janak H. Patel

University of Illinois

KLUWER ACADEMIC PUBLISHERS
Boston/Dordrecht/London

Distributors for North America:
Kluwer Academic Publishers
101 Philip Drive
Assinippi Park
Norwell, Massachusetts 02061 USA

Distributors for all others countries:
Kluwer Academic Publishers Group
Distribution Centre
Post Office Box 322
3300 AH Dordrecht, THE NETHERLANDS

Library of Congress Cataloging-in-Publication Data

Choudhary, Alok N. (Alok Nidhi), 1961-
    Parallel architectures and parallel algorithms for integrated
vision systems / by Alok N. Choudhary and Janak H. Patel.
        p.   cm. — (The Kluwer international series in engineering and
computer science : 108. Robotics)
    Includes bibliographical references and index.
    ISBN 0-7923-9078-4
    1. Computer vision.   2. Parallel processing (Electronic computers)
3. Computer architecture.   I. Patel, Janak H., 1948-
II. Title.   III. Series: Kluwer international series in engineering
and computer science : SECS 108.   IV. Series: Kluwer international
series in engineering and computer science.   Robotics.
TA1632.C47   1990
006.3'7—dc20   90-4889
                                                                                        CIP

Printed in the United States of America

Contents

List of Figures

List Of Tables

Preface

Computer vision is one of the most complex and computationally intensive problem. Like any other computationally intensive problems, parallel processing has been suggested as an approach to solving the problems in computer vision. Computer vision employs algorithms from a wide range of areas such as image and signal processing, advanced mathematics, graph theory, databases and artificial intelligence. Hence, not only are the computing requirements for solving vision problems tremendous but they also demand computers that are efficient to solve problems exhibiting vastly different characteristics.

With recent advances in VLSI design technology, Single Instruction Multiple Data (SIMD) massively parallel computers have been proposed and built. However, such architectures have been shown to be useful for solving a very limited subset of the problems in vision. Specifically, algorithms from low level vision that involve computations closely mimicking the architecture and require simple control and computations are suitable for massively parallel SIMD computers. An Integrated Vision System (IVS) involves computations from low to high level vision to be executed in a systematic fashion and repeatedly. The interaction between computations and information dependent nature of the computations suggests that architectural requirements for computer vision systems can not be satisfied by massively parallel SIMD computers.

Hierarchical and partitionable architectures for vision systems have been suggested to provide the required flexibility, reconfigurability and partionability to solve computer vision problems. The main objective in designing such architectures is that they can be adapted to perform a given computation efficiently while different computation are being performed in other parts of the architecture. Since the performance of a system is governed both by task characteristics and architecture characteristics, it is critical that an architecture can be reconfigured to match the characteristics of the task. Hierarchical architectures allow distributed control, hierarchical load balancing, and different size partitions for changing task requirements. Although, in principle, such architectures are attractive, only a limited effort has been extended towards design and development of such architectures.

This book examines issues in designing hierarchical and partionable architectures for integrated vision systems. Furthermore, important issues such as how to map different algorithm and what techniques can be used to implement algorithms with different characteristics are also addressed. In order to be able to design a good architecture, it is important to understand the characteristics of the problems for which the architecture is targeted. This book first develops a model of computation for problems in integrated vision systems. The model examines and incorporates the computation, communication and other requirements for parallel implementation of IVSs. Using these requirements, a criterion is developed to design hierarchical and partitionable multiprocessor architectures for IVSs. A multiprocessor architecture (NETRA) is discussed and is critiqued in the light of the earlier developed criterion. Techniques to map algorithms on NETRA, analysis and implementation on a simulated architecture are described.

Although several multiprocessor architectures have been designed and built, and many are commercially available, techniques to map parallel algorithms and implement algorithms to achieve good performance have lagged far behind. A user needs to have an in-depth understanding of an architecture in order to efficiently use it for parallel processing. This book develops several techniques for mapping, scheduling and load balancing of IVS algorithms. The techniques are general enough to be applied to many IVSs and can be used on any MIMD shared or distributed memory machines. This versatility is obtained by considering the characteristics of the involved tasks in IVSs and independent of the architectures on which these systems will be implemented. There is a tremendous scope to enhance and improve these techniques as well as employ them in parallelization tools and compilers. The research issues discussed in this book are discussed with reference to IVSs, they can be easily extended to a general hierarchical and partitionable architectures.

Overview of the Book

This book contains seven chapters.

Chapter 1 introduces integrated vision systems. A brief discussion on the computational complexities of such systems is presented. It contains an overview of several multiprocessor architectures and presents a critique of the architectures. A discussion on the architectures' salient features, their capabilities and limitations is presented.

Chapter 2 presents a model of computation for IVSs. The model is presented from parallel processing perspective. An attempt is made to capture the computation requirements, to recognize data dependencies between tasks, and capture the temporal flow of computation. The model is used to develop architectural requirements for multiprocessors for IVSs applications.

A hierarchical and partitionable architecture called "NETRA" is presented in Chapter 3. NETRA is a recursively defined tree-type hierarchical architecture whose leaf nodes consist of a cluster of processors connected with a programmable crossbar with selective broadcast capability to provide for desired flexibility. The scheduling processors and the cluster processor are connected to a global memory through a multistage circuit switched network. A discussion is presented that critically examines the features of NETRA in the light of architectural requirements developed in Chapter 2.

Chapter 4 presents how to map an algorithm on a processor cluster in NETRA in various modes such as SIMD and MIMD. Then performance evaluation of algorithms when mapped on one cluster is presented. The algorithms are chosen so that they exhibit different communication requirements when mapped in parallel. Performance of some algorithms on a simulated cluster is also presented.

Inter-cluster communication is discussed in Chapter 5. A general method of analysis of inter-cluster communication is presented. Two alternative inter-cluster communication networks, namely, bus and multistage, are evaluated. The analysis of inter-cluster communication is used to evaluate performance of various algorithms when mapped across multiple clusters.

Chapter 6 presents data decomposition, load balancing and task scheduling techniques for data dependent algorithms. The techniques exploit the knowledge about the data gathered from the current task and use the knowledge about involved computations in the next task in order to partition the data onto the available processors so that load balancing and high utilization are achieved. In an IVS, in most cases, such information can be available because the flow of tasks and their dependencies are known in advance. In order to evaluate the performance, implementation results for a few algorithms that are part of a motion estimation system are presented when implemented on a commercially available shared and distributed machines.

Summary, conclusions and directions for future work are presented in Chapter 7.

Acknowledgements

We would like to thank Professors Narendra Ahuja, Prithviraj Banerjee and Thomas Huang for their comments, suggestions and and invaluable contributions. We are particularly thankful to Subhodev Das and Mun Leung for providing their helpful suggestions and sharing their knowledge, which have helped us expand the scope of this work.

Special thanks go to our friends and colleagues in the Center for High Performance and Reliable Computing as well as in the Vision Group of the Coordinated Science Laboratory for their assistance and suggestions. We extend our appreciation to all the secretaries in our group for their help.

Finally, we would like to acknowledge support from National Aerospace and Space Administration for providing funding for this work. This work was supported by NASA grant NAG 1-613.

Chapter 1

Introduction

1.1. Computational Complexities in Vision

One of the most important, difficult and computationally intensive problems in the field of artificial intelligence is computer vision. There is no consensus today on the definition and scope of computer vision. The problem of artificial vision is as old as the field of computer science and engineering. Researchers have devoted much time in attempting to define and solve parts of the problems for many years. However, to say that computer vision is in its infancy today is a correct judgment of the state of the art in artificial vision. Furthermore, nobody knows the answer to the question of whether it is possible to make artificial vision as powerful and general as human vision. One of the many reasons for not knowing the answer is that little is understood about human vision itself.

There are several approaches to tackling the computational problems in computer vision. One of the approaches, which is also the oldest, is to use the computational powers of computers and their development in various fields of computer science and engineering, such as signal processing, mathematical and scientific algorithms, and graph theory. The other approach, which is relatively recent, is to somehow mimic the computations performed in the human brain. This approach is termed as the neural network approach. However, tremendous computational power in one form or another is needed in both the approaches.

Computer vision and image understanding algorithms employ a very broad spectrum of techniques from several areas such as signal processing, advanced mathematics, graph theory, and artificial intelligence. The computational requirements to perform algorithms from these fields are tremendous when executed individually, and when they need to be integrated in a meaningful way to perform a broader function in a reasonable amount of time, the computation becomes almost intractable. For example, consider interpretation of a changing scene at 30 frames per second. The amount of data to be handled per second itself is almost 25 Mbytes (million bytes) assuming a moderate resolution of 512×512 pixels per frame with each pixel of three bytes (one byte for each color and 256 grey level). The amount of computation required for simple image transformations, labeling, grouping, surface reconstruction or motion analysis is very difficult to estimate; however, for many applications it can be in the range of 100 – 10,000 billion instructions per second [1]. This is raw processing power and does not include the complexities involved in a system such as interactions among various algorithms, input-output of data, managing system resources, and fault-tolerance. Therefore, the vision problem is of tremendous interest to computer architects and it presents them with great challenges.

Having discussed the need to provide tremendous processing power in an architecture for computer vision, the next question is how can that processing power capability be provided? Parallel processing, which has progressed tremendously in the past decade, seems to be the consensus approach to providing the necessary computational power. Fortunately, most algorithms that are part of a vision system are in general, characterized by massive parallelism. For low level processing, spatial decomposition of an image provides a natural way of generating parallel tasks. For higher level analysis operations, parallelization may also be based on other image characteristics and may be data dependent. In fact, parallel processing has been suggested as the approach to provide computational power needed for most computational intensive problems such as scientific, vision or any other because technological limits are being reached in how fast a serial processor can perform. But the next question is what form of parallel processing, and what type of multiprocessor architectures are suitable for vision application? It may be easier to provide raw processing power by parallel processing, but the more important and difficult question is how to design multiprocessors so that the available processing power can be used efficiently. Since there is no consensus as to what a vision system consists of, another problem is how to evaluate or compare one architecture with another. Recently, efforts have been made to provide a framework and benchmark to evaluate multiprocessor architectures for vision which not only attempt to measure the processing power of an architecture but also test other architectural issues such as I/O, ability to perform algorithms with varying characteristics, and effect on the performance due to interactions between tasks [1].

This book attempts to identify various issues in multiprocessor architectures and parallel algorithms for computer vision. The approach is to consider the computational requirements for vision in an integrated environment rather than to propose architectural solutions to perform one or more algorithms efficiently and fast. We define the computational requirements for an

Integrated Vision System (IVS), for which there is no general definition. However, an application dependent definition of an IVS is possible. For example, object recognition, a system that takes an image (or a set of images) as input and produces an output that describes the object can be considered an IVS. However, a system (or an algorithm) that takes an image input and produces its Discrete Fourier Transform (DFT) is not considered an Integrated Vision System, though computing DFT itself may be a step or a part of an IVS. In fact, it is important to distinguish between image processing and computer vision (or IVS). Image processing involves transforming images by applying one or more algorithms to the input in order to make it more useful for interpretation by humans. For example, image enhancement, noise reduction, scaling, and thresholding constitute image processing operations. Integrated Vision System, on the other hand, involves interpretation and recognition by the system itself using input data, parameters and knowledge base without any interference from humans. That is, the system is completely an automated vision system in the ideal case. Therefore, IVS can be defined as a system which employs a subset of vision algorithms in a systematic way to produce a meaningful output. The computational requirements for such an integrated vision system are tremendous [2].

Vision algorithms are normally divided into three levels: low level, intermediate level and high level. Low level algorithms are mostly image processing algorithms. These algorithms, in general, are very regular in structure, involve data independent and local computations, and involve pixel data. Available parallelism is normally on the pixel level. Intermediate level algorithms perform computations on the output produced by low level algorithms and involve more complex data structures, data dependent algorithms, symbolic processing, and involve varying degree of parallelism which itself depends on the data and the nature of the computation. Finally, high level algorithms not only exhibit most of the properties of intermediate level algorithms but also involve top-down processing in which knowledge based interpretation is performed. Therefore, the algorithms may involve accessing databases, performing enormous searches and include other artificial intelligence algorithms.

An Integrated Vision System will normally consist of algorithms from all levels of processing. Therefore, in addition to providing tremendous raw processing power an architecture must be capable of the following. It must have the ability to transform pixel data into a set of meaningful symbols that describe it, to process pixels, symbol data and other complex data structures in parallel, and the ability to simultaneously perform low, intermediate, and high level algorithms, and fast I/O. These requirements and others mean that an architecture must be reconfigurable, provide flexible and fast communication structures between processing elements, provide different types of processing (such as SIMD, MIMD) to most efficiently execute algorithms from different levels of processing, be efficient in performing dynamic resource allocation and task scheduling, be partitionable into independent subsystems which can work on different computations simultaneously, be fault-tolerant and provide fast I/O bandwidth to keep up with a tremendous amount of data flow.

Design of multiprocessor architectures for IVSs, therefore, must address the requirements posed by the above discussed characteristics of algorithms that are part of an IVS. In this book, we present a model of computation for IVSs for parallel processing. The model attempts to capture the properties of IVS algorithms, data flow and interactions between various tasks. Our model not only captures the computation requirements presented in the Image Understanding Benchmark presented in [1] but it also provides for another dimension (time) of computation which is absent in the benchmark. Then we present an architecture for integrated vision systems called "NETRA." NETRA, in its original form, was first proposed by Sharma, Patel and Ahuja in [3]. We propose several refinements to the architecture based on our understanding of computational requirements for IVSs. An elaborate discussion is presented that gives the rationale behind the design. Several common vision algorithms are used to evaluate the performance of the architecture and alternative communication strategies. The algorithms are mapped using the multidimensional divide-and-conquer paradigm [4] which is an attractive mechanism for providing parallelism in all levels of processing.

1.2. Review of Multiprocessor Architectures

The advent of VLSI technology has enabled architects to produce high performance chips to perform specific applications. But these special purpose chips can only be used in an IVS as accelerators of specific algorithms. Another use of VLSI technology has been to create massively parallel Single Instruction Multiple Data (SIMD) processors for vision and other applications. There are also Multiple Instruction Multiple Data (MIMD) processors in which the number of processors is normally a few orders of magnitude less than that in SIMD (massively parallel) machines', however, each processor is a powerful general purpose processor with its own program and data memory. Normally, MIMD machines fall into two categories: shared memory and distributed memory machines, though many architectures exhibit both paradigms. Within these classifications, multiprocessors are distinguished according to the interconnection topology between processors or processor-memories. Finally, there are systolic arrays, hierarchical, partitionable multiprocessor architectures that have been proposed and studied for vision applications. Many of the architectures have been proposed just for vision application but almost all multiprocessors have been studied for vision applications. In the following discussion, we examine many of the architectures, describe the topology, salient features and limitations, and discuss their advantages and disadvantages with respect to solving vision problems. We use the topology as the main classification of the architectures in describing them; however, within a topology, if machines exist that can be further classified, then we will present a discussion on them.

1.2.1. Mesh connected computers

Mesh connected multiprocessors have been one of the first multiprocessors proposed for computer vision and image processing applications. For image processing applications, meshes seem to be an obvious choice because the images map quite naturally onto its structure. Figure 1.1 shows the topology of a mesh-connected computer. A typical machine consists of a large number of processing elements (PEs) arranged in a square array. Most typical is a 4-connected mesh in which each processor is connected to its four nearest neighbors. However, 6 (hexagonal) and 8 connected meshes have also been proposed. Most machines built on this topology are SIMD type of machines. Each PE has its local memory, and it responds in SIMD mode to the instructions broadcast by a controller. The PEs can be selectively masked using mask registers.

The advantage of this architecture is that images map quite naturally onto its structure. When the image size matches the size of the multiprocessor (e.g., $N \times N$ mesh for $N \times N$ image), maximum parallelism can be obtained for those operations that require computations on individual pixels or a very small neighborhood of pixels. However, this type of architecture has several limitations. There are many low and intermediate level vision algorithms that involve grouping or matching of image structures which are spatially distant in an image. But in meshes, communication across large distances is expensive and inefficient. Therefore, unless the computation is regular and local, meshes do not perform well. Furthermore, meshes have been proposed only as SIMD machines, and that means lack of MIMD processing capability that is necessary to support high level vision. In order to cost-effectively build a multiprocessor with thousands of processors, individual processors must be small, given the technological limitations. Normally, a typical machine will have PEs with 1-bit ALUs and a small memory, which may be sufficient for small pixel based operations but definitely lacks the power that is needed for intermediate and high level operations. Finally, to most efficiently use a mesh, it is required that the data size exactly match the processor size, which is a severe limitation.

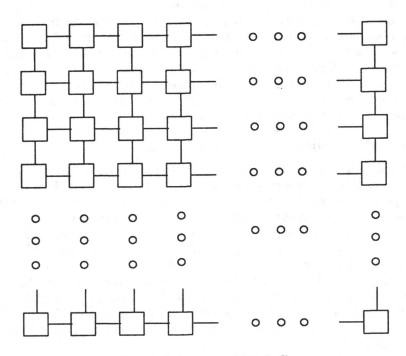

Figure 1.1 : A 4-Connected Mesh Computer

Several mesh-connected multiprocessors have been built. Examples of mesh-connected computers include CLIP-4 [5, 6, 7], GRID [8], GAPP [9], and the MPP [10, 11, 12]. Each of these machines has its own special features, but all of them have the same general form. One major drawback of these machines has been the inefficiency of collecting results and rapid evaluation of the results due to communication bottlenecks. This reflects the fact that they were designed and built as stand-alone image processors used primarily for image enhancements in which the results of processing are intended for interpretation by humans rather than forming the first stage of an autonomous vision system. Several enhancements to a mesh have been proposed to alleviate the global communication problems. Wrapped around connections of the boundary PEs is one of them in which top row PEs are connected to the bottom row PEs and the first column PEs are connected to the last column PEs. This arrangement is called Torus. This reduces the long distance communication time, but the order still remains the same. Other enhancements include connecting PEs in rows and columns by busses to broadcast common data, but these enhancements do not alter the basic structure, advantages and limitations of a mesh-connected computer.

1.2.2. Pyramid computers

The concept of pyramid computers is essentially an extension of meshes in the third dimension. This structure has been proposed in various forms, but the main idea is that an image sized mesh-connected array is augmented by layers of successively lower resolution mesh-connected arrays as shown in Figure 1.2. Each array in a pyramid is typically one fourth as large as the array below it. Except for the bottom array, each PE in a pyramid is connected to four processors in the level below it, in addition to the neighbor connections in the same level. Formally, a pyramid consists of $(1/2)\log N + 1$ levels, where the i-th level, $0 \leq i \leq (1/2)\log N$, is a mesh with $N/4^i$ PEs. Each level has connections to the level above and below, giving each internal PE 9 connections: 4 to its children in the level below, 4 to its nearest neighbors at the same level, and 1 to its parent in the level above. All the PEs operate in SIMD mode under the directions of a single controller. Several pyramids have been proposed and built and examples include PAPIA [13], SPHINX [14], MPP Pyramid [15], HCL Pyramid [16, 17], and others [18, 19, 20, 21].

Pyramid multiprocessor architecture provides straightforward implementation of the divide-and-conquer based approach. Such pyramids are natural candidates for executing divide-and-conquer algorithms, as they most closely mirror the flow of information in these algorithms. The pyramid processor provides the capability for quickly changing the resolution of an image, which can significantly improve the execution speed of some low level algorithms, especially for those that depend upon communication between cells that are spatially distant in an image. However, pyramid processors are more difficult to build than meshes because of the more complex arrangement for communication links and require twice the number of processing elements for the same image resolution. Hence, no pyramid multiprocessor has been built commercially.

Despite the fact that a pyramid machine has multiple levels of processing elements, it should not be concluded that a pyramid is suitable for implementing the multiple levels of processing required in an integrated vision system. The pyramid only implements an image resolution hierarchy, whereas vision requires an architecture that implements a hierarchy of abstraction levels. In a pyramid machine, all the processors are identical and execute in SIMD fashion. A vision machine, on the other hand, requires a different type of processing at different levels and in a variety of modes of parallelism including both SIMD and MIMD. Furthermore, from a purely

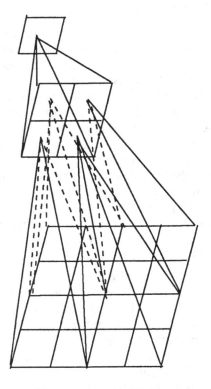

Figure 1.2 : A Pyramid Multiprocessor

architectural point of view, utilization of pyramid processors, in general, tends to be very low because at each level the slowest processor(s) is (are) the bottleneck, and the pipeline of computation (bottom-up) is limited by the slowest processor, thereby limiting the utilization at all the levels. Therefore, the pyramid machines, like meshes, are mostly suitable for early low level vision algorithm only and lack the flexibility for and processing capabilities needed for complex vision applications.

The effectiveness and performance of architectures such as pyramid, array processors, and meshes are limited as architectures for integrated vision systems due to several reasons. First, they are mostly suitable for SIMD types of algorithms which only constitute low level vision operations.

Second, the architectures are inflexible due to the rigid interconnections. Third, the number of processors needed to solve a problem of reasonable size is thousands. Such a large number of processors is not only cost prohibitive, but the processors themselves cannot be very powerful and can have only limited features due to technological limitations. Fourth, it is normally assumed that the problem size exactly matches the number of processors available. Most of the time it is not clear how to adapt algorithms so that problems of different sizes can be solved on the same number of processors. Finally, the problem of input-output of data and fault-tolerance is rarely addressed in any of these architectures. It is important to note that no matter how fast or powerful a particular architecture is, its utilization can be limited by the bandwidth of the I/O. Furthermore, due to rigidity of most architectures, a failure normally either results in the failure of the entire system, or the performance degrades tremendously. It is important that any architecture for such a complex problem should provide for graceful degradation which can be achieved by flexibility of the interconnect and capabilities to efficiently reconfigure and partition the architecture.

1.2.3. Hypercube multiprocessors

Hypercube multiprocessors provide more efficient long distance communication that is absent in meshes or pyramids. Machines in this class consist of processors connected by communication links whose arrangement is topologically equivalent to an n-dimensional cube. A hypercube consists of $N=2^n$ PEs for an n dimensional cube. Each PE is connected to n other PEs such that their binary representations differ in exactly one bit position. Therefore, any PE can communicate with any other PE using at most n communication links. Figure 1.3 illustrates the organization of a hypercube multiprocessor.

Several commercially available machines have been built that use the hypercube topology. Both SIMD and MIMD types of machines have been built. The Connection machine is an SIMD hypercube multiprocessor [22]. However, in a connection machine, two communication networks are provided. Each PE is connected to its four NEWS neighbors through a NEWS network, and groups of processors are connected in a hypercube fashion that provides efficient long distance communication. Such a machine can be used for most low level vision algorithms and some intermediate vision algorithms. However, like in other SIMD machines, lack of MIMD processing capability precludes its use for high level vision. Furthermore, low and intermediate processing cannot occur simultaneously, which is a necessity for complex, real-time vision systems.

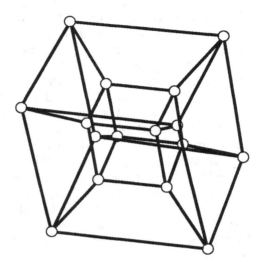

Figure 1.3 : A Hypercube Multiprocessor

MIMD hypercube multiprocessors are also commercially available. In fact, several companies have built MIMD hypercubes of large sizes (up to 1024 processors). Examples include Intel Hypercube [23], NCube [24], and Cosmic cube [25]. A typical processor node in a machine consists of a general purpose microprocessor (e.g., 80386 and coprocessor 80387 in Intel iPSC/2), local memory and routing hardware. Each multiprocessor is controlled by a host processor. The advantage of the hypercubes is that they provide efficient long distance communication between processors. Although hypercube machines with large dimensionality have been built, current systems are not very efficient due to slow communication bandwidths and tremendous overheads of running an application. However, a hypercube machine can be used for some intermediate level and high level vision applications. One major disadvantage with hypercubes is that in order to efficiently utilize the machine, the algorithms running should somehow use the underlying topology. Nevertheless, of most existing machines, hypercubes (especially MIMD) have proved to be the most cost effective machines for research and development of scientific as well as vision applications and have helped tremendously in learning issues in parallel processing in general. But we think that due to rigidity of the architecture, lack of global control and inefficient communication (especially in large machines) will prohibit hypercubes from being multiprocessors for complex vision applications.

1.2.4. Shared memory machines

Shared memory multiprocessors proposed and built are normally MIMD machines. Each PE is a general purpose processor with a small local memory. Each PE has access to a large global memory through an interconnection structure that connects the PEs and the global memory. The design of an interconnection network itself has been a huge area of research. Almost all the machines built today have variations of two common interconnection networks: bus-based and multistage interconnection networks. All the interconnections (in the machines built) are a variation of the two approaches. Bus-based systems have a limitation on the number of processors, due to the bus access bottlenecks, and therefore, are not easily scalable. However, design is relatively simple and cost-effective. Machines have been built using up to 32 processors in such a system. Sequent Balance [26] and Encore Multimax [27] are two good examples of bus-based, shared memory multiprocessors that are commercially available.

Another class of shared memory multiprocessors use multistage interconnection networks for processor-processor or processor-memory interconnections. Some bottlenecks of bus based systems are alleviated in such a system; however, the interconnection networks are complex to build. Scalability in such architectures is much better than that in bus-based systems, and machines with up to 128 processors have been built. Most machines built or being built have been for research purposes. Examples of these machines include BBN Butterfly (Commercially available) [28], IBM RP-3 [29], and Cedar, which is being developed at University of Illinois [30]. The main advantage of shared memory architecture is the ease of programming and uniform view of the system. In other words, control of information and synchronization is much easier compared to that in distributed memory systems. Therefore, this class of machine is best suited for high level vision tasks. However, since communication between processor and all the interaction between cooperative tasks is done through the global memory there are bottlenecks, and hot spots occur. Furthermore, accessing global memory is at least an order of magnitude higher than accessing local memories, and therefore, communication speed is very slow compared to computation speed. Hence, such machines are efficient for only large grain parallelism tasks which have little interactions and exhibit regular memory access patterns. Since processes interact with each other using global memory shared variables, the comparative overhead of synchronization is very high and also results in hot spots. Because the actual image processing operations execute relatively quickly when they are divided among multiple processors, the process start-up and synchronization overhead rapidly grows to dominate the processing time. Therefore, scalability is definitely a problem in any shared memory multiprocessors. It is possible to build big machines, but the return of using larger sized shared memory multiprocessors to solve a problem becomes negative beyond a certain size.

1.2.5. Systolic arrays

Originally systolic arrays were proposed for special purpose computations. A systolic array multiprocessor consists of processors connected in a certain fashion in which on each machine cycle each processor takes values from its input ports, performs the required computation, and passes the results and data onto its output ports. A systolic array can be perceived as a pipeline of a series of processing stations. Once the pipe is filled with data, all of the processing stations operate on values in parallel. Systolic array elements can be either general purpose programmable function units or special purpose fixed function units. The latter are not useful for vision applications because of their inflexibility. The primary advantage provided by a programmable systolic array is high performance for low cost. They are, however, best suited for image processing tasks, but can work well with any application that involves large arrays of data and regular computation. The main disadvantage of a systolic array is that any evaluation of processing results must wait until all the data has passed through the array. If a systolic array processes an image in one frame time, then this restriction has the effect of allowing the controlling process to make the decision and change the array's programmed functions once each frame time. In a systolic array, it is thus much more difficult for a vision system to quickly and flexibly adapt its processing strategy to the actual characteristics of an image.

The Warp Processor

CMU Warp systolic processor is an example of a programmable systolic array designed and built for scientific and image processing applications [31, 32, 33, 34, 35]. The Warp machine is a systolic array computer of linearly connected cells, each of which is a programmable processor capable of performing 10 MFLOPS. Figure 1.4 shows the organization of the Warp computer (taken from [35]). A typical Warp array includes 10 cells, though it is claimed that it can be extended if more cells are needed [35]. The Warp array consists of identical cells called Warp cells, as shown in Figure 1.4 Data flow through the array on two communication channels (X and Y). Those addresses for cells' local memories and control signals that are generated by the Interface Unit propagate down the Adr channel. The direction of Y is statically reconfigurable. For more details the reader is referred to [35]. The Warp array can be used for both fine-grain and large-grain

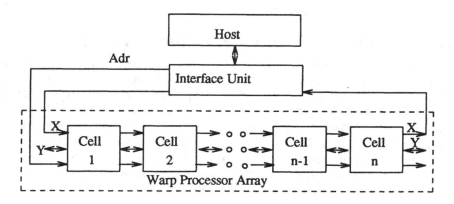

Figure 1.4 : Warp System Overview [35]

parallelism. It is efficient for fine-grained parallelism because of its high inter-cell bandwidth. It is also claimed to be efficient for large-grain parallelism because it is composed of powerful cells. Each cell is capable of operating independently; it has its own program sequencer and program memory. Even though Warp can perform in many modes, it is mostly suitable for low and intermediate level vision and does not have the desired flexibility, due to its organization, for efficiently performing high level vision algorithms.

iWarp Multiprocessor

The iWarp is an extension of the Warp multiprocessor [36]. The main additions in iWarp as compared to the Warp are; 1) iWarp has much faster processors, and 2) it is a two dimensional systolic (torus) structure that can be reconfigured as a general purpose distributed memory multiprocessor. iWarp is a system architecture for high speed signal, image and scientific computing. An iWarp cell contains a single chip processor that requires only addition of memory chips to form a complete system building block. Each iWarp component contains both a powerful computation engine (20 MFLOPS) and a high throughput (320 MBytes/sec), low latency (100-150 ns) communication engine for interfacing with other iWarp cells. iWarp is claimed to be a versatile building block for various high performance parallel systems [36]. These systems range from special purpose systolic arrays to general purpose distributed memory systems. They are also able to support both fine-grain and large-grain parallel computations simultaneously in the same system. iWarp is a product of joint effort between Carnegie Melon University and Intel Corporation.

1.2.6. Partitionable and hierarchical architectures

There have been numerous architectures designed and developed for vision that cannot be put in any of the classes discussed above. Some of the architectures include PM4 [37], PASM [38], REPLICA [39], INSPECTOR [40], and IUA [41]. Design of these architectures has addressed the issues of flexibility, partitionability, and reconfigurability which are needed in an architecture for an IVS. The following is a brief discussion on some of these architectures, their merits and limitations. An important and common characteristic of these architectures is that they are capable of being partitioned into one or more independent SIMD and MIMD subsystem.

PM4 : A Reconfigurable Multiprocessor :

The PM4 represents one of the first proposals for a reconfigurable multiprocessor capable of executing several MIMD and SIMD processes concurrently [37]. It includes a large number of processing units constituting a pool. This pool of processing units can be partitioned into groups, each one of which can operate independently in either SIMD or MIMD mode. Reconfiguration of system resources is dynamic and is primarily software controlled. The components of the system include 1) N identical Processor-Memory Units (PMUs), 2) K identical Vector Control Units (VCUs), and 3) A three-level hierarchical memory connected by a set of interconnection networks and memory managements units.

The PMUs are the basic processing units and they include a microprocessor, a local memory and a Local Memory Management Unit (LMMU). The local memory is composed of interleaved memory modules and serves as a local cache for the microprocessor. The VCUs control groups of processors operating in an SIMD mode. The Inter-Processor Communication Network (IPCN) implements permutation functions during the execution of an SIMD process. Management of the shared memory is accomplished by a Shared Memory Management Unit (SMMU), which communicates with the LMMUs and with the File Management Control Unit. The Processor-Memory Interconnection Network transfers bursts of data or instructions between the shared memory and the PMUs.

PASM :

The PASM, a Partitionable SIMD MIMD Multiprocessor, is also a dynamically reconfigurable into one or more independent SIMD and/or MIMD machines [38]. The system is composed of Parallel Computation Unit (PCU) which includes N microprocessors, N memory modules and an Interconnection Network (IN) connecting them. There is a set of Q Microcontrollers (MC), each controlling N/Q processors. Memory management tasks are distributed over a set of processors constituting the Memory Management System (MMS). The system is to operate under control of a uniprocessor System Control Unit or SCU which would be responsible for job scheduling and for coordinating loading memory modules within the PCU. Only higher levels of these tasks need to be executed on the SCU; the details can be distributed over the MCs and MMS.

Two processor-memory configurations are being examined for the PCU. In the PE-to-PE configuration, each processor has a local memory and the composite processor-memory units (PEs) communicate via the IN. When data is to be obtained from the memory of another PE, the two PEs involved cooperate to effect the transfer. Two processors are, therefore, involved for any non-local reference. In the P-to-M configuration, processors are connected on one side of the Interconnection Network and memory modules on the other. The processors do not have a local memory and can access any of the modules on the other side of the network. As a result, no explicit data transfers from one processor to another are required. However, all references now have to go via the Interconnection Network. The system can be partitioned into one or more partitions, each with RN/Q processors, where $R = 2^r$. Each partition can operate in either an SIMD or an MIMD mode. The code required to be transferred to more than one MCs (for execution on corresponding sets of N/Q processors) is broadcast to the selected MCs over a bus.

The REPLICA :

The REPLICA was designed as a special purpose computer for multisensory perception of 3-D objects [39]. Its main features include support for clean and flexible partitionability with minimal fragmentation and modularity. The machine consists of the following components: 1) A pool of N processing elements, each with a local memory, 2) A two-level control hierarchy. At level-1 are the controllers and at level-2 are monitors. The monitor layer is responsible for scheduling tasks and reconfiguring and partitioning the system. The set of M controllers is uniformly distributed over the system-one for each group of N/M processors, 3) A memory management system controlling a large shared memory and a secondary memory and 4) Four interconnection networks. One network handles communication between monitors and controllers. A second one, a capability enhanced crossbar, connects the controller to the PEs. Shift-register rings are used for communication between PEs within a partition. Finally, a high bandwidth bus is suggested for communication between the PEs, controllers, sensors and memory. This bus is to support I/O and the transfers of data and programs between main memory and the local memories of the PEs.

The clean partitionability is attributed to the capabilities of controller-processor and processor-processor interconnection networks. It is claimed that these networks allow variable size partitions composed of arbitrary subsets of processors. Partitions can be set up rapidly and are totally isolated from each other (i.e., the partitions are clean).

The Image Understanding Architecture :

The Image Understanding Architecture (IUA) integrates parallel processors operating simultaneously at three levels of computational granularity in a tightly-coupled architecture [41]. Each level of the IUA is a parallel processor that is distinctly different from the other two levels, in order to best meet the processing needs for different levels of algorithms in a computer vision application. Communication between levels takes place via parallel data and control paths.

The bottom level of the architecture contains an associative processor called the Content Addressable Array Parallel Processor (CAAPP). The CAAPP is a 512×512 array of 1-bit serial processors designed to operate on arrays of pixels and to construct intermediate-level tokens from events in an image. At the intermediate level, an array of 64×64 16-bit processors, called the Intermediate Communications and Associative Processor (ICAP), are used for the intermediate level of processing. Specifically, the processors are used for retrieving, comparing and matching tokens, computing geometric relationships between tokens, and constructing new tokens that describe more abstract entities. At the top level (called high level) is the Symbolic Processing Array (SPA) which is a set of 64 processors capable of executing LISP programs. Their function is to support computation involving inference, hypothesis generation and verification, analysis of uncertainty, model-based processing and control of processing at the lower levels. Currently, a 1/64th of the IUA is currently being constructed by the University of Massachusetts and Hughes Research Laboratories.

1.3. Organization

This book contains 7 chapters. The following is an overview of the contents of each chapter and the organization of this book.

Chapter 2 presents a model of computation for IVSs. The model is presented from parallel processing perspective. An attempt is made to capture the computation requirements, to recognize data dependencies between tasks, and capture the temporal flow of computation. The model is used to develop architectural requirements for multiprocessors for IVSs applications. These requirements broadly describe features that should be present in a multiprocessor design in order for it to be efficient for IVSs.

Architecture of NETRA is presented in Chapter 3. NETRA is a recursively defined tree-type hierarchical architecture whose leaf nodes consist of a cluster of processors connected with a programmable crossbar with selective broadcast capability to provide for desired flexibility. The internal nodes are scheduling processors and their function is task scheduling, load balancing, and global memory management. All the scheduling processors and the cluster processor are connected to a global memory through a multistage circuit switched network. The processors in clusters can operate in SIMD, MIMD or systolic mode, and therefore, suitable for both low level as well as high level vision algorithms. A discussion is presented that critically examines the features of NETRA in the light of architectural requirements developed in Chapter 2.

Chapter 4 presents how to map an algorithm on a processor cluster in NETRA in various modes such as SIMD and MIMD. Then performance evaluation of algorithms when mapped on one cluster is presented. The algorithms are chosen so that they exhibit different communication requirements when mapped in parallel. Performance of some algorithms on a simulated cluster is also presented. It is concluded that good speedups and performance can be obtained on a cluster because of the availability of a programmable crossbar which provides the necessary flexibility in mapping algorithms with varying characteristics.

Inter-cluster communication is discussed in Chapter 5. A general method of analysis of inter-cluster communication is presented. Two alternative inter-cluster communication networks, namely, bus and multistage, are evaluated. The analysis of inter-cluster communication is used to evaluate performance of various algorithms when mapped across multiple clusters. When algorithms are mapped onto multiple clusters, the performance is affected by conflicts and interference in the global interconnection networks. These effects are incorporated in the analysis, and it is concluded that if interconnection bandwidth is fast enough then good performance results can be obtained even in the presence of high conflicts.

Chapter 6 presents data decomposition, load balancing and task scheduling techniques for data dependent algorithms. The techniques exploit the knowledge about the data gathered from the current task and use the knowledge about involved computations in the next task in order to partition the data onto the available processors so that load balancing and high utilization are achieved. In an IVS, in most cases, such information can be available because the flow of tasks and their dependencies are known in advance. In order to evaluate the performance, implementation results for a few algorithms that are part of a motion estimation system are presented when implemented on a hypercube multiprocessor system. The reason for choosing a hypercube multiprocessor is that using an existing machine helps capture the overheads associated with such techniques.

Summary, conclusions and directions for future work are presented in Chapter 7.

Chapter 2

Model of Computation

Computer vision transcends a wide range of representations and forms of processing. Despite advances in many sub-areas of computer vision, there is no consensus on a unified approach to vision. However, one can define certain general characteristics of an Integrated Vision System (IVS) from computational perspective. For example, it is known that a "vision system" must be able to perform diverse sets of complex operations on a massive amount of data at high speeds. Motion sequences at moderate resolution (512×512 pixels) and typical frame rate (30 frames/sec) in color (3 bytes) involve more than 20 Mbytes of data per second. The amount of computation required for dynamic scene interpretation including labeling objects, surface reconstruction and motion analysis is difficult to estimate; however, for many applications computational power in the range of 10^{12-14} instructions per second is required [41]. Not only are the raw processing needs tremendous, but varying the type of processing capabilities (such as number crunching, symbol manipulation, and data processing) are required.

Parallel processing in some form has been accepted as the approach to providing the necessary computational power to solve complex vision problems. But several questions remain. What type of parallel processing is best suited? What architectural features are needed? How is the performance of a multiprocessor architecture measured and how is its effectiveness as an architecture for IVSs evaluated? Several attempts have been made to define benchmarks that capture processing needs for vision tasks [42, 43, 44]. Recently Weems et al. designed a benchmark for integrated vision systems that attempts to capture different forms of processing, and includes algorithms with different characteristics and their interactions [1]. However, the benchmark does not include "time" dimension in the sense that motion and time varying information are omitted from the benchmark.

In this chapter, we define a model of computation for integrated vision systems (IVS) from parallel processing perspective. The model also includes the time dimension and is more general. It can be used to critically examine a multiprocessor architecture proposed for IVSs. However, it is not a benchmark that can be used to evaluate architectures. Using the model we attempt to identify the architecture requirements for IVSs as well as provide a framework to design new benchmarks to evaluate architectures.

2.1. Parallelism in IVSs

Available parallelism in integrated vision systems can be placed in two broad categories: namely, *Spatial* and *Temporal Parallelism*. Within the categories, the available parallelism can be further sub-divided into different classes. The classes depend on the type of tasks (or algorithms) constituting the system, the type of architectures on which the tasks are to be implemented, the methodology used to implement tasks, interactions between the tasks, and control and data flow between the tasks. For example, a task may exhibit suitability for data parallelism at the lowest level and can be implemented on a massively parallel SIMD architecture; or a task may exhibit data dependent, non-uniform computation, and therefore, be suitable for implementation on an MIMD architecture in a sub-tasks parallelism mode in which sub-tasks cooperate to produce results.

Spatial Parallelism is one in which similar operations are applied in all parts of the image data. That is, the data can be divided into many granules and distributed to subtasks which may execute on different processors in parallel. Most vision algorithms exhibit this type of parallelism. In an IVS, each task operates on the output data of the previous task in the system. Therefore, the type of data, and data structures may be different for each task in the system but each form of data can be partitioned into several granules to be processed in parallel. For example, consider an IVS that performs object recognition. The input image is smoothed using some filtering operation, then on the smoothed image an operator is applied for feature extraction, features with similar characteristics are grouped, then matching with the models is performed. Each of these tasks takes the output of the previous tasks as its input and produces an output which becomes the input for the next task. Note that within spatial parallelism, depending on the computation involved, an algorithm implementation may be suitable for data parallelism or task parallelism or both.

Temporal Parallelism is available when these tasks are repeated on a time sequence of images or on different resolutions of images. For example, the system in which motion of a moving object is estimated takes a sequence of images of the moving object and performs the same set of computation on all image frame(s). The processing of each frame or a set of frames can be done in parallel with the processing of frames of other time instances.

Figure 2.1 shows the computational model for IVS which illustrates the above mentioned characteristics of an IVS. Each pipeline shows a number of tasks applied to a set of inputs. The input to the first task in a pipeline is the image, and the input to the rest of the tasks is the output of the previous task. The set of pipelines illustrates that the entire pipeline of tasks is repeated on different images in time and/or resolution. Each block in the pipeline represents one task. Each task is decomposed into subtasks to be performed in parallel. For example, T_1 is one task, and $T_1(d_1)$ is a subtask of T_1 operating on data granule d_1. The figure shows m tasks in the pipeline. The number of subtasks depends on the amount of data in a granule and number of available processors. $D_{i,i+1}$ represents data transfer from task T_i to task T_{i+1} in the pipeline. The model does not make any assumptions about a particular implementation of a task.

2.2. Data Dependencies

Existence of spatial and temporal parallelism may also result in two types of data dependencies, namely, *spatial data dependency* and *temporal data dependency*. Spatial data dependency can be classified into intratask data dependency and intertask data dependency. Intratask data dependencies arise when a set of subtasks needs to exchange data in order to execute a task in parallel. The exchange of data may be needed during the execution of the algorithm, or to combine the partial results, or both. Therefore, each task itself is a collection of subtasks which may be represented as a graph with nodes representing the subtasks and edges representing communication between subtasks. Intertask data dependency denotes the transfer and reorganization of data to be passed onto the next task in the pipeline. The mode of communication may be subtasks of the current tasks to the subtasks of the next task, or collection and reorganization of the output data of the current task and then redistribution of the data for the next task. The choice and method depend on the underlying parallel architecture and mapping of algorithms. Temporal data dependency is similar to spatial data dependency except that some form of output generated by tasks executed on the previous image frames may be needed by one or more tasks executing on the current image frames. A simple example of such a dependency is the IVS of motion estimation in which features from the previous image frames are needed in the processing of the current image frames so that features can be matched to establish correspondence between features of different time frames.

The total computation to execute one pipeline includes time to input data, time to output data and results, sum of the times to execute all tasks in the pipeline (which includes computation time of subtasks and communication time between subtasks) and, data transfer and reorganization time

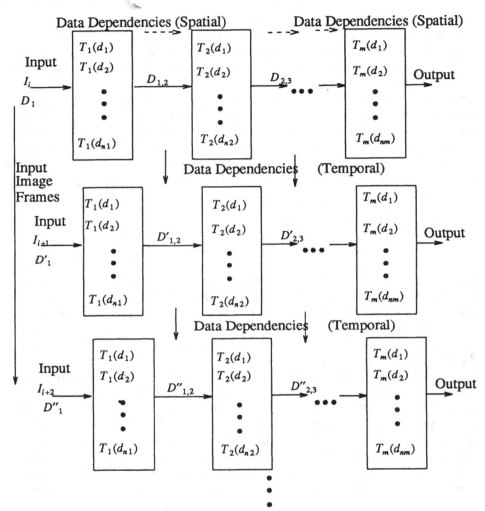

Figure 2.1 : Model of Computation for an Integrated Vision System

between two successive tasks. Let's denote t_{cp} as computation time for a subtask, t_{comm} as total communication time for a task, t_{in} as data input time, t_{out} as data output time, and t_d as data transfer and reorganization time. Then time to complete task i, denoted as τ_i is given by

$$\tau_i = \underset{1 \leq j \leq ni}{MAX} \, t_{cp}(T_i(d_j)) + t_{comm}(T_i) \tag{2.1}$$

Total time to execute one pipeline including the input and output of data is given by

$$t_{tot} = \sum_{i=1}^{i=m} \tau_i + \sum_{i=1}^{i=m-1} t_d(D_{i,i+1}) + t_{in} + t_{out} \tag{2.2}$$

Let us now consider some characteristics of the algorithms involved in IVS, and using the above model determine desired features and capabilities of a multiprocessor architecture suitable for IVS. First, an IVS involves algorithms from all levels of processing, i.e., an IVS normally includes low, medium and high level vision algorithms. Typically, the first few tasks of the pipeline are low level algorithms and the last few are high level algorithms. The low level algorithms are well understood and well defined. They are normally data independent, have regular structure, and spatial parallelism is mostly available at pixel level. They are well suited for both SIMD and MIMD type of processing. If communication between processors is fast enough, almost linear speedups are possible. Therefore, an architecture for IVS should be capable of efficiently executing low level algorithms and algorithms suited for SIMD type of processing. Also, data I/O should not be a bottleneck because otherwise, speedups through parallelism can be nullified. Examples of low level algorithms include most transforms, filtering algorithms, and convolution algorithms.

High level algorithms are not well understood. They are normally global data dependent, involve more complex data structures (compared to pixel representation), and need varying communication for parallel processing. These type of algorithms are more suited for MIMD type of processing. Hence, the architecture should be capable of executing MIMD algorithms efficiently.

2.3. Features and Capabilities of Parallel Architectures for IVSs

The following are the architecture requirements for a multiprocessor architecture to be suitable for integrated vision systems. First, the ability to transform image data (pixel data) into a set of meaningful symbols that describe it. Second, the ability to process pixel and symbol data in parallel as well as concurrently. Third, the ability to simultaneously maintain low, intermediate and high level representations, and the ability to perform low, intermediate and high level algorithms simultaneously on inter-related or independent data. Fourth, fast I/O and processing rates for huge amounts of data at all levels of computations. Fifth, the ability to select particular subsets of data for varying types of processing. Finally, the ability to perform top-down as well as bottom-up processing efficiently, and the ability to report the results efficiently. These are some of the broad requirements for an architecture for integrated vision systems. From the above discussion we can transform the requirements into specific architecture requirements as presented below.

(1) Reconfigurability: From the model and the preceding discussion it is clear that multiple levels of representations and stages of processing are essential and require very different types of processing. Hence, the architecture should be capable of executing both SIMD and MIMD type computations efficiently. That is, it should be possible to reconfigure the architecture such that each algorithm can be implemented efficiently using the most suited mode of computation.

(2) Flexible Communication: Fine grained and high speed communication is required both among the processes at each level and between the different processing levels. The communication requirements vary for different algorithms. The communication pattern between processors executing subtasks of a larger task depends on the algorithm involved in the task. If the connectivity between processors is too rigid then the communication overhead of intratask and intertask communication may become prohibitive. Therefore, it is desirable that the communication be flexible in order to provide the most efficient communication with low overhead.

(3) Resource Allocation and Partitionability: As we discussed earlier, there are several tasks with vastly different characteristics and computational requirements in an IVS. These tasks need to exist simultaneously in the system. Therefore, the system should be partitionable into many independently controlled subsystems to execute each task. Since the high level algorithms exhibit varying level of parallelism and data dependent performance, it should be possible to allocate resources (such as processors, memory) dynamically to meet the performance requirements.

(4) Load Balancing and Task Scheduling: Load balancing and task scheduling are very important, especially for high level vision algorithms, which are data dependent, and therefore, in order to obtain better utilization of resources and better speedups, dividing the computation equally among the processor is critical [45]. The underlying architecture on which load balancing is done and the type of algorithm(s) involved contribute significantly to how well load

balancing can be achieved. In low level algorithms since the computations are data independent, partitioning data equally among the processors normally balances the load among them. However, for high level algorithms, more sophisticated load balancing and scheduling strategies are needed. The architecture should include features such that it is easy to perform load balancing and task scheduling and that the overhead of doing so is minimal.

(5)   Topology and Data Size Independent Mapping: For a system as complex as an IVS, if the underlying architecture and its interconnect is rigid such that the problem size that can be solved on it or how it can be mapped is tied to the interconnection, the effectiveness of the architecture will diminish as an architecture for an IVS.

(6)   Fault-Tolerance: Fault-tolerance is an important part of a system of such complexity. A failure in a processor or communication structure should not affect the performance drastically, which is normally the case when rigid interconnections are present between processors. The architecture should provide for graceful degradation in case of failures.

(7)   Input-Output: It is most often the case that an architecture is able to perform very well on some algorithms, and high speedups are obtained, but input-output (I/O) of data is inefficient. I/O is an integral part of a system and if it is a bottleneck then performance of the system will be limited.

2.4. Examples of Integrated Vision Systems

2.4.1. Image understanding benchmark system

Recently, a DARPA sponsored effort has been directed towards developing benchmarks to evaluate architectures for integrated vision systems, and the benchmarks and rationale behind it appeared in [1]. We will briefly present a discussion on the benchmark as it partly represents an integrated vision systems. The following are some of its features. The benchmark involves a simple image domain with well-defined, well-behaved objects. It requires both bottom-up (data-directed) and top-down (knowledge or model-directed) processing. The top-down processing can involve processing of low and intermediate level data to extract additional features from the data, or can involve control of low and intermediate level processes to reduce the total amount of computation required. It tests low level operations such as convolution, thresholding, connected component labeling, edge tracking, median filter, hough transform, convex hull, and corner detection. It requires utilization of information from two sensors in order to complete the interpretation process. It tests grouping operations and graph matching, as representative examples of intermediate level and high level processing, respectively. It requires use of both integer and floating point representations. Finally, it tests the communication channels between symbolic and numeric levels of processing. The reader is referred to [1] for more details.

The benchmark includes most characteristics of a typical integrated vision system, or at least is a good representation of what type of processing may be needed in such a system. However, it does not span the entire spectrum. Most important, the benchmark does not include motion information. That is, it does not capture the real-time image input or time-varying information processing. Our model, presented earlier in this chapter, tries to capture the time varying characteristics of an integrated system and captures most characteristics of the integrated benchmark presented in [1]. Furthermore, it provides a framework to develop benchmarks in the future even though several refinements in the model need to be performed and a more detailed view has to be provided.

2.4.2. Motion estimation and object recognition

In this system, sequence of images of a scene containing moving object(s) is used to compute the motion of the object(s) in the scene, and using the motion parameters and features from the images, object recognition is performed [46]. The computation involves extracting zero crossings (convolution, template matching and thresholding), stereo matching (graph matching and grouping) of features, hough transform, and model directed object recognition in which features obtained from the image data are correlated with the features of the set of model objects in order to obtain the best match. Therefore, this system also involves both bottom-up and top-down processing. Figure 2.2 shows the computational flow for the motion estimation system in which stereo images (L_{im}, R_{im}) at each time frame are used as the input to the system. Briefly, the involved tasks in this system are as follows. The first algorithm in the system is computation of zero crossings in the images (edge detection (L_{zc} and R_{zc})). The zero crossings are used to perform stereo match between the two images of the same time frame. The stereo match algorithm provides points to compute 3-D information about the object in the scene. Using these matched points (L_{sm} and R_{sm}), the corresponding points in the image in the next time frame (L_{tm}) are located, and this task is performed by time match algorithm. Again, stereo match is used to obtain the corresponding 3-D points in the next image frame. These two sets of points provide information to compute the motion parameters. Using the motion parameters and the information from the models, object recognition module performs the task of picking the best match between

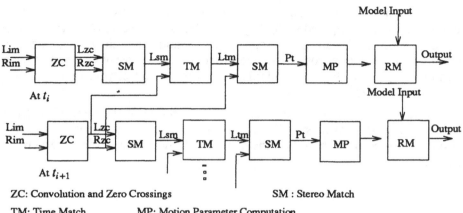

ZC: Convolution and Zero Crossings   SM : Stereo Match

TM: Time Match   MP: Motion Parameter Computation

RM: Recognition Module

Figure 2.2 : Computation Flow for Motion Estimation

models and the information from image data. The above process is repeated for each new set of input image frames.

Hence, such a system, in addition to exhibiting many of the properties included in the image understanding benchmark, also captures motion information. Therefore, several architecture features pertaining to real-time processing, and fast I/O processing can be evaluated using such a system.

The following describes the computations involvde in various steps in the motion estimation system [46]. A detailed description of the involved computations is included in order to understand the characteristics of such algorithms. The motion estimation algorithm consists of two processes. The first process is feature points extraction. Since the feature points used in the algorithms are edge points, we can extract them by locating the zero crossings of an image. The second process is matching and has three subprocesses which are i) stereo matching, ii) time matching and iii) elimination of multiple matches. The basic evidences exploited in these subprocesses to obtain unambiguous matched point pairs are the normalized correlation coefficient and the zero crossing patterns [47].

Feature points

The feature points used in this algorithm are zero crossing points of an image which are computed using Laplacian-Gaussian masks [48]. In order to eliminate non-significant zero crossing points and maintain enough details, we threshold the zero crossing image based on the intensity gradient at each zero crossing point. Figure 2.4 depicts the thresholded zero crossing images of the pictures shown in Figure 2.3.

Each zero crossing point is associated with one of the sixteen possible zero crossing patterns as suggested in [47]. The similarity between any two zero crossing points is based on the directional difference of their zero crossing patterns.

Figure 2.3 : Stereo Image Pairs at t_7 and t_8

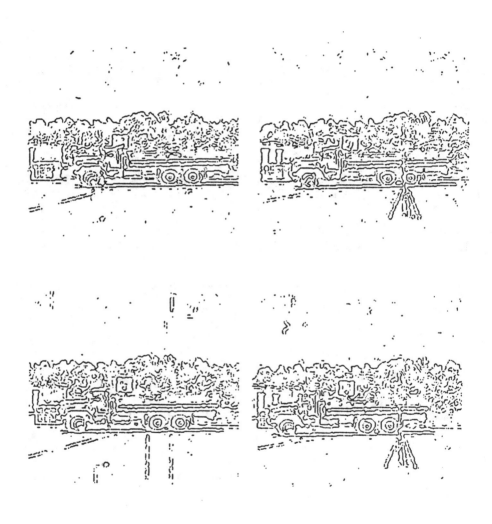

Figure 2.4 : Zero Crossings of the Images in Figure 2.3

Matching

Once zero crossings are extracted in all the involved images, the matching process is applied to find point correspondences among the images (two stereo image pairs at two consecutive time instants, i. e., t_{i-1} and t_i). The evidences used in this process to obtain matched point pairs are the normalized correlation coefficient and the directional difference weight as mentioned above. Furthermore, in order to limit the search space, the heuristic of limited displacement or disparity between frames is exploited. The matching processes in motion estimation consist of six steps described below.

1) Perform stereo (from left to right) matching in the t_{i-1} stereo image pair.

2) Obtain unambiguous matched point pairs by eliminating multiple matches.

3) Perform time matching between the unambiguous matched points in the left t_{i-1} image and the feature points of the left t_i image.

4) Obtain unambiguous matched point pairs from the time matched points by eliminating multiple time matches.

5) Perform stereo matching between the unambiguous matched points (obtained in step (4)) in the left t_i image and the feature points of the right t_i image.

6) Obtain unambiguous matched point pairs from the results of t_i stereo matching by eliminating multiple matches.

The results of the above steps are two sets of unambiguous stereo matched point pairs at time instant t_{i-1} and t_i. These two sets are related through Steps (3) and (4), the matching over time. Therefore, all the unambiguous matched points that correspond to each other among the two stereo image pairs at time instants t_{i-1} and t_i can be selected.

Stereo matching

This is the subprocess to obtain the matched point in the right image for each matchable zero crossing point in the corresponding left image of the same stereo pair. Since the imaging setup is the parallel axis method, the epipolar line constraint is exploited in solving the stereo matching problem. As a result, we have a one-dimensional search space instead of a two-dimensional search space in the stereo matching process. A typical search space in the right image for a matchable zero crossing point in the left image is on the left side of the transferred location of that particular left image zero crossing point. However, by using the heuristic of limited disparity between frames, the search space is limited [46].

Let S_{rl} be the set of all non-horizontal zero crossing points in the right image within the search space of a zero crossing point in the left image. The stereo matching process is as follows :

For each point in S_{rl},

i)   Calculate the normalized correlation coefficient with a template size of $s \times s$ between the grey level images of left and right at the corresponding locations.

ii)   If the normalized correlation value is less a threshold value, discard that particular point in the search space in the remaining steps.

iii) Calculate the directional difference weight between the left and the right zero crossing point (within the search space).

iv)   Obtain the total weight as the combination of the correlation coefficient and the directional difference weight.

v)   Among all elements of S_{rl}, the point with the maximum total weight is considered as the matched point for the corresponding zero crossing point in the left image.

Time matching

This subprocess is used to obtain the matched point in the left t_i image for each candidate zero crossing point in the corresponding left t_{i-1} image. Similar to the stereo matching process, the heuristic of limited displacement (instead of disparity) between frames is exploited in solving the time matching problem. It is assumed that the total motion between the t_{i-1} and t_i frames is within f pixels in the vertical direction and h pixels (from right to left) in the horizontal direction. Hence, the search space for each candidate zero crossing point in the left t_{i-1} image is a window of size $f \times h$ pixels on the left side of its transferred location in the left t_i image. Any zero crossing point (except horizontal ones) inside this window is a potential match point for the corresponding candidate zero crossing point in the left t_{i-1} image. The time matching process is outlined below.

For each non-horizontal zero crossing point in the left t_i image within the search space of a zero crossing point in the left t_{i-1} image,

i) Calculate the normalized correlation coefficient with a template size of $t \times t$ between the grey level image of t_{i-1} and t_i at the corresponding locations.

ii) If the normalized correlation coefficient ρ_t is less than a threshold value, discard that particular point in the remaining steps.

iii) Calculate the directional difference weight between the left t_{i-1} and the left t_i zero crossing point (within the search space).

iv) Obtain the total weight as the combination of the correlation coefficient and the directional difference weight.

v) Within a search window in the left t_i image, the zero crossing point with the maximum total weight w_t value is considered as the match point for the corresponding zero crossing point in the left t_{i-1} image.

Elimination of multiple matches

After the subprocess of either stereo matching or time matching, there may be multiple matches for some zero crossing points in either the left image or the left t_{i-1} image. Same procedure is used in eliminating both types (stereo or time) of multiple matches, except different sizes of the search window are exploited. For stereo matching, a 1-D search window is used on the left side of a multiple matched point. On the other hand, for time matching, a two-dimensional search window on the left side of a multiple matched point is used. The remaining steps of the procedure for determining unambiguous matched points are as follows :

i) At the position of a multiple match in either the right image for stereo matching or the left t_i image for time matching, open a search window either with a size of d_{max} or with a size of $2f \times h$, respectively.

ii) Within the search window, locate all the positions that have the same (multiple) match.

iii) If all the positions are within the neighborhood region, calculate the total weight. The position with the highest value in its total weight is regarded as the correct match.

iv)   If one or more positions are outside the neighborhood region , the following disambiguation procedure described in the following section to resolve the multiple matches.

v)   If multiple matches still exist after the application of the above steps, they all are discarded from the match set.

Disambiguation

This procedure is used only if step (iv) in the above discussion is true. In this procedure, the neighboring unambiguous matched points around a multiple matched point are used as one of the supporting evidences in determining the correct match. The other evidences used are the normalized correlation coefficient and the directional difference weight. The steps are as follows :

i)   At each position, calculate the normalized correlation coefficient and the directional difference weight.

ii)   Assign a correlation coefficient rank and zero crossing pattern rank to each position according to its normalized correlation coefficient and directional difference weight. The position with the highest value in normalized correlation coefficient or directional difference weight has the highest rank.

iii)   At each position, check for unambiguous matched neighbors. If it has two attached unambiguous matched neighbors, a neighbor weight of 3 is assigned. On the other hand, if it has only one attached unambiguous matched neighbor, a neighbor weight of 2 is assigned.

iv)   At each position, open a check window of size 5×5 but exclude the center 3×3 region and count the number of unambiguous matched points.

v)   At each position, calculate the total possibility as the sum of the ranks, the weight and the number. The position with the highest value is considered as the correct match.

Figure 2.5 shows all the unambigous points obtained after performing all but steps "MP" and "RM" step in figure 2.2. For details of all the steps the reader is referred to [46].

Both examples of IVSs described in this chapter map well into the computation model proposed earlier. All the steps of motion estimation and object recognition system fit into the model. Furthermore, all the interactions between tasks are captured by the model. Several refinements and additions to the model are possible depending on what level of interaction and information is to be incorporated. For system level description the model is sufficient to describe an IVS.

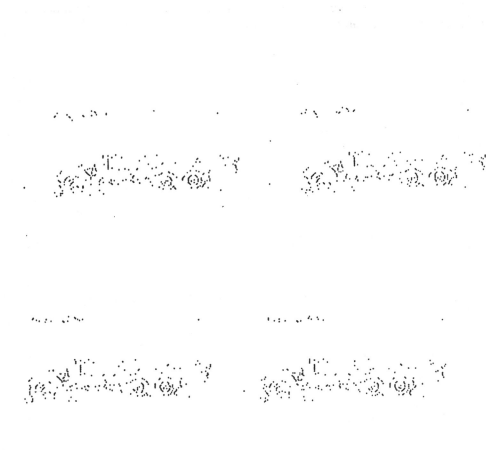

Figure 2.5 : Unambiguous Matched Points of the Images in Figure 2.3

Chapter 3

Architecture of NETRA

This chapter contains a detailed description of the architecture of NETRA. The first four sections describe the components of NETRA, their functions, capabilities and features. The last section critically examines the architecture in view of the computational requirements for IVS developed in the previous chapter.

Figure 3.1 shows the architecture of "NETRA," which is a recursively defined hierarchical multiprocessor system and provides distributed as well as shared memory environment. The architecture consists of the following components :

(1)   A large number (1000 - 10000) of *Processing Elements (PEs)*, organized into clusters of 16 to 64 PEs each.

(2)   A tree of *Distributing-and-Scheduling-Processors (DSPs)* that make up the task distribution and control structure of the multiprocessor.

(3)   A parallel pipelined shared *Global Memory* and a *Global Interconnection* that links the PEs and DSPs to the Global Memory.

3.1. Processor Clusters

The clusters consist of 16 to 64 PEs, each with its own program and data memory. Each PE is a general purpose processor with a high speed floating point capability. They form a layer below the DSP-tree, with a leaf DSP associated with each cluster. PEs within a cluster also share a common data memory. The PEs, the DSP associated with the cluster, and the shared memory are connected together with a crossbar switch. The crossbar switch permits point-to-point communications as well as selective broadcast by the DSP or any of the PEs. Figure 3.2 shows the cluster organization. A 4x4 crossbar is shown as an example of the implementation of the crossbar switch. The crossbar design consists of pass transistors connecting the input and output data lines. The switches are controlled by control bits indicating the connection pattern. If a processor or DSP needs to broadcast, then all the

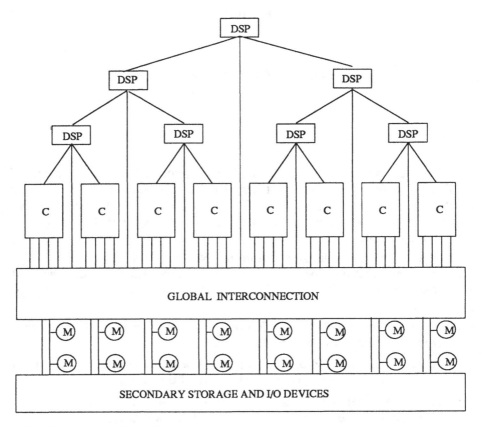

DSP : Distributing and Scheduling Processor

C : Processor Cluster   M : Memory Module

Figure 3.1 : Organization of NETRA

control bits in its row are made one. In order to connect processor P_i to processor P_j, control bit (i,j) is set to one, and the rest of the control bits in row i and column j are off.

Clusters can operate in SIMD mode, systolic mode, or MIMD mode. In an SIMD mode, PEs in a cluster execute identical instruction streams from private memories in a lock-step fashion. In systolic mode, PEs repetitively execute an instruction or set of instructions on data streams from one or more PEs. In both cases, communication between PEs is synchronous. In MIMD

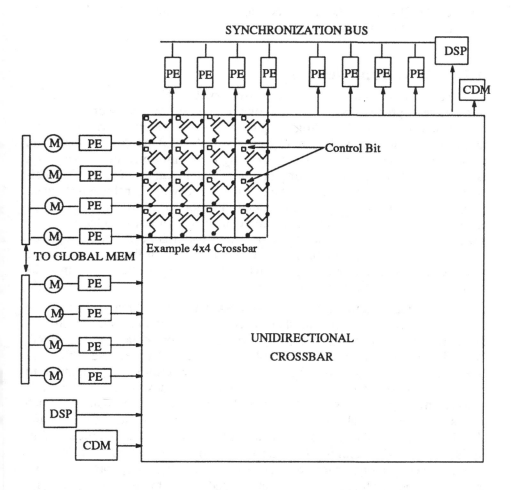

SYNCHRONIZATION BUS

Example 4x4 Crossbar

TO GLOBAL MEM

Control Bit

UNIDIRECTIONAL
CROSSBAR

PE : PROCESSOR   M : LOCAL MEMORY

CDM : COMMON DATA MEMORY

Figure 3.2 : Organization of Processor Cluster

mode, PEs asynchronously execute instruction streams resident in their private memories. The streams may not be identical. In order to synchronize the processors in a cluster, a synchronization bus is provided which is used by processors to indicate to the DSP that a processor(s) has finished its computation or a processor wants to change the communication pattern. The DSP can either poll the processors or the processors can interrupt the DSP using the synchronization bus.

3.1.1. Crossbar design

There is no arbitration in the crossbar switch. That is, the interconnection between processors has to be programmed before processors can communicate with each other. Programming a crossbar requires writing a communication pattern into the control memory of the crossbar. A processor can alter the communication pattern by updating the control memory as long as it does not conflict with the existing communication pattern. The DSP associated with a cluster can write into the control memory to alter the communication pattern. The most common communication patterns, such as linear arrays, trees, meshes, pyramids, shuffle-exchanges, cubes, broadcast, can be stored in the memory of the crossbar. These patterns need not be supplied externally. Therefore, switching to a different pattern in the crossbar can be fast because switching only requires writing the patterns into the control bits of the crossbar switches from its control memory.

The advantages of such a crossbar design are the following: First, since there is no arbitration, the crossbar is relatively faster than one which provides arbitration because switching and arbitration delays are avoided. Second, it is easier to design and implement the crossbar because arbitration is absent, and therefore, switches are simple. Furthermore, it is possible to implement systolic algorithms using the crossbar because it can transfer data at the same or greater speed than required by the systolic computation. Such a crossbar is easily scalable. Unlike other interconnections, such as cubes and shuffle-exchanges, the scalability need not be in power of 2. A unit scalability is possible. Furthermore, for the same reason, it is easy to provide fault-tolerance because one spare processor can replace any failed processor, and one extra crossbar link can replace any failed link. This is possible because there is no inherent structure that connects the processor and each processor, (link) is topologically equivalent to any other processor (link).

3.1.2. Scalability of crossbar

Figure 3.3a) depicts a 1 bit 4×4 crossbar switch. In order to obtain byte or word parallel crossbar, the crossbar switches can be stacked together as shown in Figure 3.3b). The control, address and communication pattern information is exactly the same in all the stacked switches. Figures 3.3c), d) and e) illustrate the size scalability. Figure 3.3c) shows how a 4×8 crossbar can be obtained from two 4×4 crossbars. Similarly, Figures 3.3d) and e) illustrate how 8×4 and 8×8 crossbars can be obtained, respectively. Note that the smallest switch need not be a bit crossbar. Depending on the technology and availability of the I/O pins, it can be of any size (such as 4 bit or a byte). Furthermore, depending on the available pins, it can be a 16×16 or 32×32 bit crossbar. Finally, sizes of the crossbar need not be a multiple of two but can be any arbitrary.

3.2. The DSP Hierarchy

The DSP-tree is an n-ary tree with nodes corresponding to DSPs and edges to bi-directional communication links. Each DSP node is composed of a processor, a buffer memory, and a corresponding controller.

The tree structure has two primary functions. First, it represents the control hierarchy for the multiprocessor. A DSP serves as a controller for the subtree structure under it. Each task starts at a node on an appropriate level in the tree, and is recursively distributed at each level of the subtree under the node. At the bottom of the tree, the subtasks are executed on a processor cluster in the desired mode (SIMD or MIMD) and under the supervision of the leaf DSP.

The second function is that of distributing the programs to leaf DSPs and the PEs. Vision algorithms are characterized by a large number of identical parallel processes that exploit the spatial parallelism and operate on different data sets. It would be highly wasteful if each PE issued a separate request for its copy of the program block to the global memory because it would result in unnecessary traffic through the interconnection network. Under the DSP-hierarchy approach, one copy of the program is fetched by the controlling DSP (the DSP at the root of the task subtree) and then broadcast down the subtree to the selected PEs. Also, DSP hierarchy provides communication paths between clusters to transfer control information or data from one cluster to others. Finally, the DSP-tree is responsible for Global Memory management.

3.3. Global Memory

The multiport global memory is a parallel-pipelined structure as introduced in [49]. Given a memory(chip)-access-time of T processor-cycles, each line has T memory modules. It accepts a request in each cycle and responds after a delay of T cycles. Since an L-port memory has L lines, the memory can support a bandwidth of L words per cycle.

Data and programs are organized in memory in *blocks*. Blocks correspond to "units" of data and programs. The size of a block is variable and is determined by the underlying tasks and their data structures and data

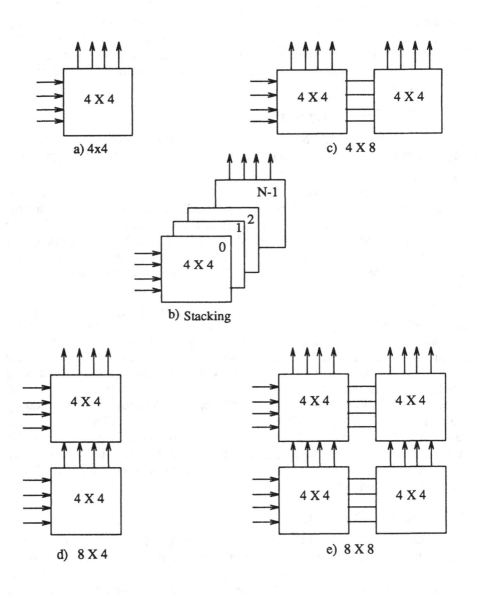

Figure 3.3 : Scalability of Crossbar

requirements. A large number of blocks may together constitute an entire program or an entire image. Memory requests are made for blocks. The PEs and DSPs are connected to the global memory with a global interconnection network.

The global memory is capable of queuing requests made for blocks that have not yet been written into. Each line (or port) has a Memory-line Controller (MLC) which maintains a list of read requests to the line and services them when the block arrives. It maintains a table of *tokens* corresponding to blocks on the line, together with their length, virtual address and *full/empty* status. The MLC is also responsible for virtual memory management functions.

Two main functions of the global memory are input-output of data and program to and from the DSPs and processor clusters, and to provide inter-cluster communication between various tasks as well as within a task if a task is mapped onto more than one cluster.

3.4. Global Interconnection

Currently, two alternative global interconnection schemes are being evaluated. First is a high speed bus which is connected to one port from each cluster and to the global memory. The second is a multistage interconnection network connecting the global memory and cluster processors.

3.4.1. Interconnection network

The PEs and the DSPs are connected to the Global Memory using a multistage circuit-switching interconnection network. Data is transferred through the network in pages. A page is transferred from the global memory to the processors which is given in the header as a destination port address and the header also contains the starting address of the page in the global memory. When the data is written into the global memory, only the starting address needs to be stated. In each case, end-of-page may be indicated using an extra flag bit appended to each word.

3.4.2. Global bus

We are evaluating an alternative strategy to connect DSPs, clusters and the global memory using a high speed bus. In this organization one port of each cluster will be connected to the high speed bus. Also, each DSP will be connected to the bus. Processors that need to communicate with processors in other clusters use explicit messages to send and receive data from the other processors. Figure 3.4 illustrates this method. A processor P_i in cluster C_i can send data to a processor P_j in cluster C_j as shown in the Figure. P_i sends the data to the DSP_i, which sends the data to DSP_j in a burst mode. DSP_j then sends the data to the processor P_j. We are evaluating both alternatives for intercluster communication.

3.5. IVS Computation Requirements and NETRA

In the following discussion we examine the architecture in the light of requirements for an IVS discussed in Chapter 2.

Reconfigurability (Computation Modes)

The clusters in NETRA provide SIMD, MIMD and systolic capabilities. As we discussed earlier, it is desirable to have these modes of operations in a multiprocessor system for IVS so that all levels of algorithms can be executed efficiently. For example, consider matrix multiplication operation. We will show how it can be performed in SIMD and systolic modes. Assume that the computation requires obtaining matrix $C = A \times B$. For simplicity, also assume that the cluster size is P and the matrix dimensions are $P \times P$. Note that this assumption is made to simplify the example description. In general, any arbitrary size computation can be performed independent of the data or cluster size.

GLOBAL BUS

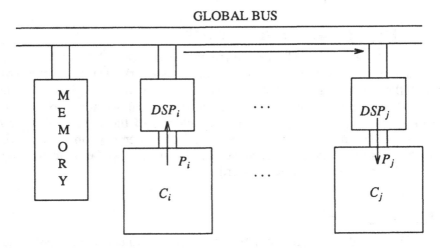

Figure 3.4 : An Alternative Strategy for Inter-Cluster Communication

SIMD Mode

The algorithm can be mapped as follows. Each processor is assigned a column of the B matrix, i.e., processor P_i is assigned column B_i. Then the DSP broadcasts each row to the cluster processor, which computes the inner products of the rows with their corresponding columns in lock-step fashion. Note that the elements of the A matrix can be continuously broadcast by DSP, row by row without any interruptions, and therefore, efficient pipelining of data input, multiply, and accumulate operations can be achieved. Figure 3.5a) illustrates a SIMD configuration of a cluster. The following pseudo code describes the DSP and processor (P_k's program, $0 \leq k \leq P-1$) program.

SIMD Computation

DSP	P_k
1. *FOR i=0 to i=P-1 DO*	1. -
2. *connect(DSP,P_i)*	2. -
3. *out(column B_i)*	3. *in(column B_i)*
4. *END_FOR*	4. -
5. *connect(DSP, all)*	5. -
6. *FOR i=0 to i=P-1 DO*	6. $c_{ik} = 0$
7. *FOR j=0 to j=P-1 DO*	7. *FOR j=0 to j=P-1 DO*
8. *out(a_{ij})*	8. *in(a_{ij})*
9. *END_FOR*	9. $c_{ik} = c_{ik} + a_{ij}*b_{jk}$
10. *END_FOR*	10. *END_FOR*

In the above code, the computation proceeds as follows. In the first three lines, the DSP connects with each processor through the crossbar and writes the column on the output port. That column is input by the corresponding processor. In statement 5, the DSP connects with all the processors in a broadcast mode. Then, from statement 6 onward, the DSP broadcasts the data from matrix A in row major order, and each processor computes the inner product with each row. Finally, each processor has a column of the output matrix. It should be mentioned that the above code describes the operation in principle and does not exactly depict the timing of operations.

Systolic Mode

The same computation can be performed in a systolic mode. The DSP can reconfigure the cluster in a circular linear array after distributing columns of matrix B to processors as before. Then DSP assigns row A_i of matrix A to processor P_i. Each processor computes the inner product of its row with its column and at the same time writes the element of the row on the outout port. This element of the row is input to the next processor. Therefore, each processor receives the rows of matrix A in a systolic fashion, and the computation is performed in the systolic fashion. Note that the computation and communication can be efficiently pipelined. In the code, it is depicted by statements 7-10. Each element of the row is used by a processor and immediately written onto the output port, and at the same time, the processor receives an element of the row of the previous processor. Therefore, every P cycles a processor computes new element of the C matrix from the new rows it receives every P cycles. Again, note that the code describes only the logic of the computation and does not include the timing information. Figure 3.5b) illustrates a systolic configuration of a cluster.

Partitioning and Resource Allocation

There are several tasks with vastly different characteristics in an IVS, and therefore, the number of processors needed for each task may be different and may be needed in different computational modes. Hence, partitionability and dynamic resource allocations are keys to high performance. Much effort has been devoted towards investigating the partitionability of interconnection networks [50, 51]. Approaches such as in [38, 51] are, however, required only when processes are tightly coupled. In the above case physical partitions are established. In other words, links are reserved for specific point-to-point communication while a process executes. Whenever a new process is to be instantiated, required resources should be free and linked together in a specified manner. The partitioning is, in effect, isolated from the rest of the system.

Systolic Computation

DSP	P_k
1. *FOR i=0 to i=P-1 DO*	*1.* -
2. *connect(DSP,P_i)*	*2.* -
3. *out(column B_i)*	*3. in(column B_i)*
4. *out(row A_i)*	*4. in(column A_i)*
5. *END_FOR*	*5.* -
6. *connect(P_i to P_{i+1} mod P)*	*6. $c_{ii}=0$*
7. -	*7. FOR j=0 to j=P-1 DO*
8. -	*8. $c_{ii} = c_{ii} + a_{ij}*b_{ji}$*
9. -	*9. out(a_{ij}), in(a_{i-1j})*
10. -	*10. END_FOR*
11. -	*11. repeat 7-10 for each row*

Partitioning in NETRA is achieved as follows. When a task is to be allocated, the set of subtrees of DSPs is identified such that the required number of PEs is available at their leaves. One of the subtrees is chosen on the basis of characteristics of the task, locality constraints and load balancing considerations. The chosen DSP represents the root of the control hierarchy for the task. Together with the DSPs in its subtree, it manages the execution of the task. Note that partitioning is only virtual. The PEs are not required to be physically isolated from the rest of the system. Once the subtree is chosen, the processes may execute in SIMD, MIMD or systolic mode. The following are some of the advantages of such a scheme. First, only one copy of the programs needs to be fetched, thereby reducing the traffic through the global interconnection network. Second, simple load balancing techniques may be employed while allocating tasks. The tasks of global memory management can be distributed over the DSP tree by assigning it to the DSP at the root of the subtree executing the subtask. Finally, locality is maintained within the control hierarchy, which limits the intratask communication to within the subtree.

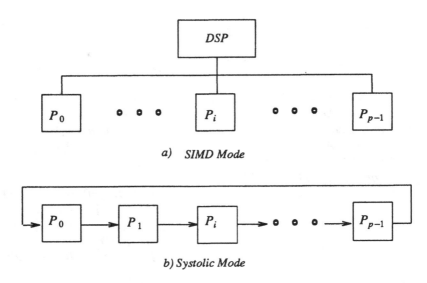

a)   *SIMD Mode*

b) *Systolic Mode*

Figure 3.5 : An Example of SIMD and Systolic Modes of Computation
        in a Cluster

Load Balancing and Task Scheduling

Two levels of load balancing need to be employed, namely, global load
balancing and local load balancing. Global load balancing aids in partitioning and allocating the resources for tasks as discussed earlier. Local load
balancing is used to distribute computations (data) to processors executing
subtasks of a larger task. Local load balancing can be either static or
dynamic or a combination of both. With static load balancing, given a task,
its associated data and the number of processors allocated for the task, the
data is partitioned in such a way that each processor gets an equal or a comparable amount of computation [45]. In dynamic load balancing, the subtasks are dynamically assigned to the processors as and when they finish the
previously assigned tasks. In NETRA, when a task is assigned to a subtree,
the DSPs involved perform the local load balancing functions.

Using the information from local load balancing and other measures of
computation, global load balancing can be achieved hierarchically by using
the DSP hierarchy. In this scheme, each controller DSP sends its measure of
load to its parent DSP and the root DSP receives the load information for the
entire system. The root DSP then broadcasts the measure of load of the entire

system to the DPSs. When a task is to be allocated, these measures can be used to select a subtree for its execution as follows: If any subtree corresponding to the child of the current DSP has an adequate number of processors then the task is transferred to a child DSP with the lowest load; else if the current subtree has enough resources and the load is not significantly greater than the average system load then the task is allocated to the current subtree; else the current DSP transfers the task to the parent DSP.

Flexible Communication

Availability of flexible communication is critical to achieving high performance. For example, when a partition operates in SIMD mode there is a need to broadcast the programs. When a partition operates in MIMD mode, where processors in the partition cooperate in the execution of a task, one or more programs need to be transferred to the local memories of the processors. Performing the above justifies a need for selective broadcast capability. In order to take advantage of spatial parallelism in vision tasks, processors working on neighboring data need to communicate fast among themselves for high performance. The programmability and flexibility of the crossbar provide fast local communication. Most common vision algorithms, such as FFTs, filtering, convolution, counting, and transforms need a broad range of processor connectivities for efficient execution. These connectivities include arrays, pipelines, several systolic configurations, shuffle-exchanges, cubes, meshes, and pyramids. Each of these connectivities may perform well for some tasks and badly for others. Therefore, using a crossbar with a selective broadcast capability, any of the above configurations can be achieved, and consequently, optimal performance can be achieved at the clusters.

Several techniques for implementing reconfigurability between a set of PEs were studied [50, 52]. It was discovered that using a crossbar switch to connect all PEs was simpler than any other schemes. The popular argument that crossbar switches are expensive was easily thwarted. When designing communication networks in VLSI, the primary constraint is the number of pins and not the chip area. The number of pins is governed by the number of ports on the network and is independent of the type of network. Furthermore, it was realized that a crossbar with a selective broadcast capability was not only a very powerful and flexible structure, but was also simpler, scalable, and less expensive.

The need for global communication is relatively low and infrequent. Global communication is needed for intertask communication, i.e., from one task to another in the IVS pipeline. It is also needed to input and output data, to transfer data within a subsystem when a task is executed on more than one cluster, and finally, it is needed to load the programs. The most important issue in global communication is that the network speed should be matched with the crossbar speed as well as with the processors speed. The global communication is performed through the global memory using the interconnection network, or using the DSP hierarchy. Another alternative we consider is connecting all the clusters and DSPs to a global bus. Since the DSPs perform most control functions and loading of programs and data, the responsibility of intertask communication does not lie with the DSP hierarchy.

I/O and Global Memory Access

The global memory is equally accessible from all the processors and DSPs in the system. Input-Output of data from (to) sensors and other I/O devices is performed through the global memory. Since the global memory lies in the address space of each processor and provides a uniform view across the architecture, I/O is uniformly distributed. Therefore, there are no I/O bottlenecks in the system. Furthermore, the global memory provides a uniform access to the shared database, which may contain models and other system data.

A large system such as NETRA implies a large memory and a large interconnection network. Therefore, the response times to memory requests can be large and variable in a nondeterministic manner, due to conflicts in accessing global memory and interconnection network. Hence, there is a need for the PEs to be able to issue multiple requests in advance and accept out-of-order requests.

NETRA is a multiprogrammed system with a large number of processes active at any time. A process becomes active when a token corresponding to the process is entered into the Active Queue of a PE (for an MIMD process) or a cluster (for a SIMD process). Data requests for the required input data blocks are immediately issued. When all input data blocks for a process are available, it is transferred to the Ready Queue. However, while these requests are serviced, the PEs continue to execute process already in their Ready Queue. Access to memory for one process is thus overlapped with execution of another. Multitasking at the PE level, therefore, permits each PE to tolerate large and undeterministic memory access latencies. Since the assumption is that such a system will be executing an integrated vision system, as we observed in Chapter 2, there will be enough processes available and active all the time. Furthermore, the future tasks will be somewhat predictable because order of the tasks is known from the model of computation and control flow of the system.

3.6. Comparison of NETRA with Other Architectures

The following discussion presents a comparison of NETRA with other architecture proposals. Some parts of the discussion in this section have been presented in [3].

Partitionability

PM4 and PASM support partitions that contain one or more of a group of processors. Each group has a fixed size. These systems are, therefore, likely to suffer from "internal fragmentation," that is, when the number of PEs to be used is not a multiple of the number present within each group, some processors within one or more groups will remain idle, and processor utilization will suffer. REPLICA supports partitions of any size by allowing each group to execute more than one SIMD and MIMD processes. This capability is by virtue of a capability-enhanced crossbar switch used to connect the PEs. NETRA, too, provides a similar crossbar at the cluster level and is, hence able to support independent partitions within the cluster. This features eliminates internal fragmentation and improves processor utilization.

Setup of Partitions

The second aspect of partitionability is that while NETRA establishes *virtual* partitions, physical partitions are established in the other cases. NETRA does not partition the pool of processors into isolated subsystems. Instead, it merely allocates processes to subsets of processors. The process of partitioning involves only placing the token for a process in the appropriate queue. In the case of SIMD process, the tokens are placed on the active queues of DSPs controlling the selected clusters. For MIMD processes, the tokens are placed in the active queues of the selected PEs. SIMD processes are initiated by the controlling DSP. MIMD processes are, however, started when the token reaches the head of the ready queue within the PE. If the MIMD processes need to synchronize in any way, they must be started and synchronized by the cluster-DSP.

Multiprogramming

NETRA is a multiprogrammed system and is most efficient if a large number of tasks are active at all times. Multitasking at the PE level has been used as a tool to decouple system performance from large memory access latencies. As explained earlier, long delays for memory access are inherent in a large system, and the system has to be able to tolerate such delays without loss of performance.

An intelligent memory system has been employed to support multitasking. This feature essentially permits the PEs to execute queued tasks without having to associate with the data-fetching operation. The above scheme is unique to NETRA, and its absence in any other proposals makes them far less suitable for large systems.

Virtual Versus Physical Partitions

In the case of PASM, PM4 and REPLICA, communication links are dedicated to partitions; consequently, all subsystems are isolated from each other. MIMD processes, especially executing intermediate and high level algorithms, exhibit widely varying execution times since the amount of processing is data dependent. Therefore, when rigid partitions are used, processors would have to wait until all complete processing before they start executing another process. If the deviation in the processing times is great, such a waiting will reduce the processor utilization tremendously.

A second factor in favor of virtual partitioning is that it allows for easy allocation of tasks that are dynamically created. An arbitrary number of such processes may be generated, and their scheduling would be a difficult problem in all of the above proposals but NETRA. On NETRA these processes can be allocated as they are generated on the basis of load balancing and locality considerations alone. How the other systems would handle dynamically created tasks is not clear, but it would certainly require several global considerations. Since tasks would already have been scheduled to execute on specific partitions, where the dynamically created processes would fit in would have to be determined. The network would be able to support only selected partitions, and scheduling on the fly would be difficult.

Scheduling and Load Balancing

Fragmentation can be minimized in PM4, PASM and REPLICA only if scheduling is static or done considerably in advance of execution. This is because scheduling would involve global considerations such as partitionability of the interconnection network and availability of resources. A closer look, however, reveals a major difficulty.

For efficient scheduling and preloading, the scheduler should know in advance what resources will be available at a given time. In other words, it should know when each process is to end. Clearly, this requires a considerable amount of determinism in the behavior of a process. While such information is easily available for SIMD tasks, it may be impossible to obtain it for most MIMD tasks.

An essential implication is that processes cannot always be prescheduled. This in turns implies that data and programs for these processes cannot be prefetched. Consequently, processor utilization suffers drastically.

The scheduling scheme on NETRA is clearly simple and superior. The scheduling process (executing on a DSP) is not confronted with a large volume of information. It needs to consider only the average load on the sub-tree below it and the overall average load of the system. As soon as a process is placed in an active queue, a request for the input data blocks required is issued. Prescheduling and data prefetching are thus easily accomplished. A hierarchical control-structure (the DSP-tree) and simple scheduling and load balancing heuristic are, therefore, able to provide for high performance.

Chapter 4

Parallel Algorithms on a Cluster

There are two main considerations in the mapping of parallel algorithms onto a cluster. First is selection of a computation mode such as SIMD, MIMD or systolic, and the second is the number of available processors on a cluster and selection of the best way to map the algorithm. For a data dependent algorithm, there may be need for nonuniform data partitioning and local load balancing. The load balancing scheme may be static or dynamic. In the static scheme, the DSP in a cluster allocates tasks to the processors using some *a priori* knowledge about the computation such that each processor receives an average amount of computation. Under the dynamic load balancing scheme the DSP maintains a queue of ready tasks and assigns the tasks to the available processors as they become free to execute the next task.

The methodology we use for mapping parallel algorithms is multidimensional, divide-and-conquer with medium to large grain parallelism. An individual task (in the following discussion task and algorithm are used interchangeably) can be efficiently mapped using spatial parallelism, because most of the vision algorithms are performed on two dimensional data. However, integration of tasks involves exploiting both spatial as well as temporal parallelism can be exploited by recognizing intertask data dependencies.

The purpose of this chapter is to evaluate performance of several common vision algorithms when mapped onto a processor cluster. The discussion identifies the computation modes suitable for an algorithm, and suggests alternatives to map an algorithm. Furthermore, performance evaluation of each algorithm is presented using accurate analysis, and the analytical results of some of the algorithms are compared with the implementation results. It is shown that analytical results are very close to the implementation results. The analysis provides the flexibility to vary several parameters, and therefore, it is easier to study the effects of alternative approaches.

This chapter is organized as follows. Section 4.1 presents a classification of some common vision algorithms based on their computation and communication requirements. Section 4.2 briefly outlines alternative mapping strategies on a processor cluster. Section 4.3 contains mappings and analytical performance results for one algorithm from each class and discusses alternative mappings for some algorithms. Parallel implementation of some of the algorithms is presented in Section 4.4 and the results are compared with analytical results. Performance of two algorithms from Image Understanding Benchmark developed by Weems et al. [1] is also presented.

4.1. Classification of Common Vision Algorithms

We can classify some of the common vision algorithms according to their communication requirements when mapped onto parallel processors. The classification provides an insight into the performance of an algorithm depending on its communication requirements.

(1)   *Local Fixed* - In these algorithms, the output depends on a small neighborhood of input data in which the neighborhood size is normally fixed. Sobel edge detection, image scaling, and thresholding are examples of such algorithms. Figure 4.1a) illustrates that the output at point (x,y) depends on fixed size neighborhood values.

(2)   *Local Varying* - Like the local fixed algorithms, the output at each point depends on a small neighborhood of input data. However, the neighborhood size is an input parameter and is independent of the input image size. Convolutions, edge detection and most other filtering and smoothing operations are examples of such algorithms. Local varying is depicted by Figure 4.1b) in which it is shown that the output at point (x,y) depends on a varying size neighborhood which is normally an input parameter.

(3)   *Global Fixed* - In such algorithms each output point depends on the entire input image. However, the computation is normally input data independent (i.e., computation does not vary with the type of image and only depends on the size of the image). Two Dimensional Discrete Fourier Transform and Histogram computation are examples of such algorithms. Figure 4.1c) illustrates that output at a point (x,y) is dependent on all the input data points.

(4)   *Global Varying* - Unlike global fixed algorithms, in these algorithms the amount of computation and communication depends on the image input as well as its size. That is, the output may depend on the entire image or may depend on a part of the image. In other words, the computation is data dependent. Hough Transform, Connected and Component Labeling are examples of such algorithms. For example, in an image, a connected component may span only a small region, or in the

worst case the entire image may be one connected component (a spiral). Similarly, in case of hough transform, assuming that we are looking for lines, a line may span across the image (meaning its votes must come from distant pixels or edges) or it may be localized. Figure 4.1d) shows that the output of an algorithm may depend on global data, and the computation is input data dependent.

4.2. Issues in Mapping an Algorithm

Mapping a task on one cluster implies that intratask communication will only involve communication between processors of the same clusters. Figure 4.2 shows how a parallel algorithm is mapped on a cluster. Assume that there are P processors in a cluster. As shown in Figure 4.2, first program and data are loaded onto the processor cluster. Both in the case of SIMD or MIMD mode, the program is broadcast onto the cluster processors. The data division depends on the particular algorithm. If an algorithm is mapped in SIMD or systolic mode, then the compute and communication cycles will be

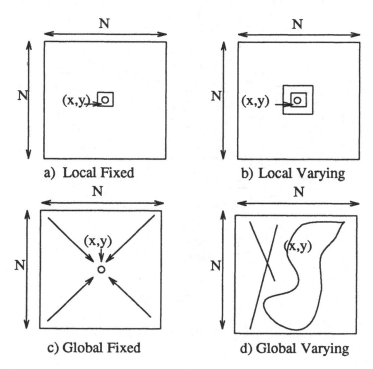

a) Local Fixed   b) Local Varying

c) Global Fixed   d) Global Varying

Figure 4.1 : Classification of Common Vision Algorithms

intermixed. If an algorithm is mapped in MIMD mode, then each processor computes its partial results and then communicates with others to exchange or merge data.

The total processing time in such a mapping consists of the following components. Program load time onto the cluster processors (t_{pl}), data load and partitioning time (t_{dl}), computation time of the divided subtasks on the processors (t_{cp}) which is the sum of the maximum processing time on a processor P_i and intra-cluster communication time (t_{comm}), and the result report time (t_{rr}). t_{dl} consist of three components: 1) data read time from the global memory (t_r) by the cluster DSP, 2) crossbar switch setup time (t_{sw}), and 3) the data broadcast and distribution time onto the cluster processors (t_{br}). The total processing time $\tau(P)$ of the parallel algorithm is given by

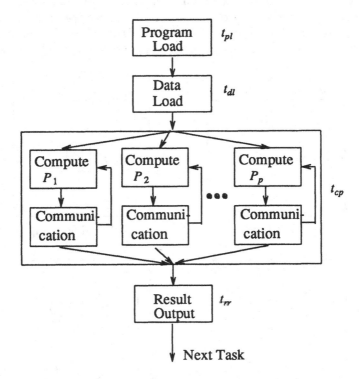

Figure 4.2 : Mapping Algorithms on One Cluster

$$\tau(P) = t_{pl} + t_{dl} + t_{cp} + t_{rr} \tag{4.1}$$

where,

$$t_{dl} = t_r + t_{setup} + t_{br} \tag{4.2}$$

and if the computation and communication do not overlap, then

$$t_{cp} = \underset{1 \le i \le P}{MAX} (t_{Pi}) + t_{comm} \tag{4.3}$$

else if computation and communication can be completely overlapped, then

$$t_{cp} = MAX \left(\underset{1 \le i \le P}{MAX} (t_{Pi}) , t_{comm} \right) \tag{4.4}$$

In the above equations, t_r depends on the effective bandwidth of the global interconnection network.

4.3. Performance Evaluation of Parallel Algorithms

In the following we illustrate how algorithms can be mapped in SIMD, systolic, and MIMD modes onto a cluster; and show how algorithms from different classes can be mapped onto the cluster. In the evaluation we discuss the computation, communication and storage requirements for the algorithms.

Table 4.1 shows the parameters used for performance evaluation. These parameters are used for all the analysis and implementation unless specified otherwise.

Table 4.1 : Parameters for Performance Evaluation

Total No. of Processors N_p	512
Cluster Size P_c	8-128
No. of Processors/Port P_p	4
Image Size $N \times N$	512 X 512
Memory Modules M	128
Processor Speed	5 MIPS, 5 MFLOPS
Network Speed (Block Transfer)	20 Mbytes/Sec.

4.3.1. 2-D convolution

2-D Convolution is a local varying type of algorithm. A 2-D convolution of an NxN image $I(i,j)$, $0<=i,j<=N$, with a kernel $W(i,j)$, $0<=i,j<=w$, can be expressed as follows :

$$G(i,j) = \sum_{m=j-w/2}^{m=j+w/2} \sum_{n=i-w/2}^{n=i+w/2} I(n,m)*W((i+w/2-n) \bmod w, (j+w/2-n) \bmod w)$$

In other words, each point in the output is replaced by a weighted sum of a window wxw around it.

The approach is to reduce 2-D convolution to a 1-D convolution with efficiency 1, i.e., without incurring additional steps. This mapping will illustrate how to map algorithms in SIMD and systolic modes on a processor cluster when the number of processors is much smaller than the problem size. Figure 4.3 shows a cluster of 64 processors. The interconnection between processors shows an abstract representation of all the connections required to perform the convolution operation. However, all the connections are not needed at the same time. We shall observe that only one input and one output connection is sufficient at any time, and that the flexibility of the crossbar can be used to obtain all the desired interconnections efficiently.

Each pixel is logically mapped onto a separate processor (as if there were as many processors available as there are pixels). Actually the image is folded and multiple pixels are mapped onto one processor. The image is folded in two dimensions in a wrap around fashion, both left to right, and top to bottom. For a cluster size P, (assume $P = pxp$), each processor has $M = N^2/P$ pixels in its local memory. In general, pixel (i,j) ; $0 \le i \le N-1$, $0 \le j \le N-1$ is mapped to processor $((i \bmod p), (j \bmod p))$. Therefore, this mapping preserves the adjacency of any two pixels even though the image is folded.

Figure 4.3 shows the flow of the distribution of data for window size 5×5. A small window is embedded in a larger one, and therefore, the same connections can be used for a larger window size with the addition of new connections for extra steps. The algorithm performs the convolution by each processor distributing its pixel values to the neighborhood in a pipelined manner.

In the following algorithm, North, South, East and West neighbors are defined in wrapped around fashion. At any step all the processors have the same neighbor connection. Figure 4.3 shows how processor (3,3)'s values will be distributed. All the processors follow the same pattern. For a processor P(i,j), N,S,E,W neighbors are defined as follows. Note that the following definition is only a logical definition, and it represents the pixel adjacency.

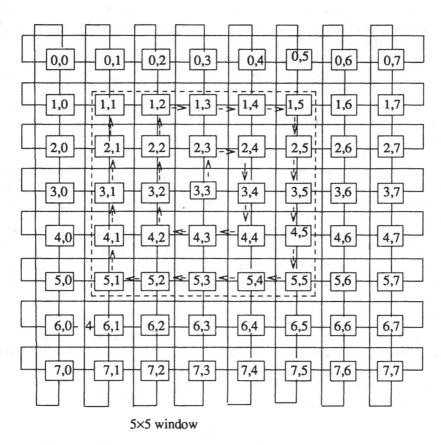

5×5 window

Figure 4.3 : Mapping on the Cluster for Convolution

The following definition does not imply any physical connections between processors.

$$N = ((i-1),j), \text{ if } (i-j) < 0, \text{ then } N = ((i-1+p), j)$$
$$S = ((i+1) \bmod p, j)$$
$$E = (i, (j+1) \bmod p)$$
$$W = (i, (j-1)), \text{ if } (j-1) < 0, \text{ then } W = (i, (j-1+p))$$

The algorithm works as follows (Figure 4.4): The DSP broadcasts the convolution weights to all the processors. Each processor multiplies its M pixels with the central weight value. In Figure 4.4 the data values at each processor are stored in a linear array and subscript (i,j) means the data value i in the connection number j. The intermediate values are stored in the running variable for each of the M pixels. The image is then shifted in a spiral manner (as shown in Figure 4.3). If the image is shifted North then the processors now multiply the pixel values with the South weight. This process is repeated w^2-1 time , i.e., for each weight. We make the following observations. First, the mapping is independent of problem or cluster size. That is, this mapping will work for all problem sizes. Second, the number of times the interconnection needs to be changed only depends on the convolution kernel size. Furthermore, at any time only one input and one output connection is required. By storing the connection patterns in the crossbar memory the switching time is negligible. Third, it is possible to overlap the computation and communication by writing the pixel to the output port as soon as it is multiplied by the appropriate weight in the current processor. The above algorithm illustrates that SIMD algorithms can be mapped efficiently onto the processor clusters using the flexibility and programmability of the interconnection.

The computation time decreases as the number of processors increases. The communication time per pixel only depends on the kernel size. The following formulae present the computation and communication times in terms of multiplication and addition operations. The factor t_{fl} denotes the floating point speed of a processors in terms of its normal instruction execution speed.

$$t_{cp} = 2 \times t_{fl} \times \left\lceil \frac{N^2}{P} \right\rceil \times w^2$$

$$t_{comm} = \left\lceil \frac{N^2}{P} \right\rceil \times w^2$$

$$t_{sw} = w^2 - 1$$

$$t_{tot} = \text{Max}(t_{cp}, t_{comm} + t_{sw})$$

Figure 4.5 shows the performance of the 2D convolution on a processor cluster. The processing time has been computed assuming a 2 MFLOP

ALGORITHM CONVOLUTION
  All the processors work in SIMD lock-step fashion.
  DSP broadcasts the convolution kernel.
  Set up Connection_array of size *wxw* in the crossbar memory by choosing.
  first *wxw* connections from the set.
  {N,E,S,S,W,W,N,N,N,E,E,E,S,S,S,W,W,W,W,N,N,N,N,E,..}.

$$M := \left\lceil \frac{N^2}{P} \right\rceil$$

  For i = 1 to M do (in parallel)
      Result(i) := $w_{i,i}$ * *data* (*i*)
  End_For

  For j = 1 to *wxw* do (in parallel)
    Set up appropriate connections on the crossbar as follows.
        connection(j) := connection_array(j)
    For i = 1 to M do (in parallel)
        Send data (pixels) on the output port to the connected
        neighbor.
        At the same time receive data from its input port.
        Result(i) := *Result* (*i*) + $w_{i,j}$ * *data* (*i,j*)
    End_For
  End_For
END CONVOLUTION

Figure 4.4 : 2-D Convolution

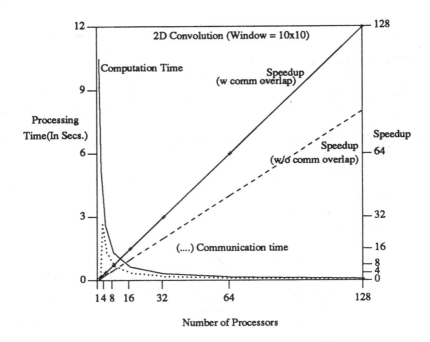

Figure 4.5 : Performance of 2D Convolution on a Processor Cluster

processor. The Figure shows two speedup graphs, one with communication overlap and the other with additive communication. The computation time decreases linearly as the number of processors increases. The total communication time per processor also decreases linearly but the communication time per pixel computation remains constant. The important observation one can make is that it is essential that the communication and computation overlap in order to obtain linear speedups. However, if the interconnection speed is not matched with the computation speed, then overlap will not be possible. Having a fast crossbar without arbitration delays provides the necessary communication speed to obtain linear speedups. Note that since computation and communication can overlap, this mapping also illustrates how systolic algorithms can be mapped.

4.3.2. Separable convolution

Separable Convolution is a two-dimensional convolution broken into two one dimensional convolution. For applications such as computation of zero crossings, separable convolution performs well [48]. The main advantage of separable convolution is that the computation requirements per pixel are reduced from $2w^2$ to $4w$. We show how it can be mapped on a cluster. This example also illustrates how an algorithm can be mapped in MIMD mode on a cluster.

The data is decomposed among the processors as follows. Each processor is assigned N/P rows of the data. Processor P_i gets rows $(i-1) \times N/P$ to $i \times N/P - 1$. Each processor computes convolution along the rows using a window of size w. Once processor P_i finishes convolution along the rows, it needs rows $(i-1) \times N/P - w/2$ to $(i-1) \times N/P - 1$, from processor P_{i-1}, and similarly, it needs the bottom $w/2$ rows from $i \times N/P$ to $i \times N/P + w/2 - 1$ from processor P_{i+1}. Therefore, a processor needs to communicate with only two processors to obtain the desired intermediate data. The boundary processors P_0 and P_{P-1} only need to communicate with one other processor. Note that if the granule size with each processor is less than $w/2$ (i.e., $N/P < w/2$), then the processors need to exchange data with number of processors given below by t_{sw}. Now, each processor computes convolution along the columns in its granule. The following are computational and communication requirements of the algorithm.

$$t_{cp} = \frac{t_{fl} \times N^2 \times 4 \times (w/2 + 1)}{P}$$

$$t_{comm} = 2 \times N \times w$$

$$T_{sw} = \left\lceil \frac{w \times P}{N} \right\rceil$$

The amount of computation per pixel in separable convolution is a function of w for a $w \times w$ kernel unlike in 2D convolution where it is a function of w^2. The amount of communication in separable convolution is fixed as shown in Figure 4.6. Therefore, the speedup is not as much as in the case of 2D convolution. There are two reasons for smaller speedup. First, the communication is not decomposable as a function of number of processors because each processor needs to exchange $w/2$ rows of intermediate results with two adjacent processors. Secondly, since the computation per pixel itself is small, the communication overhead as a fraction of computation time is large.

4.3.3. Two-dimensional FFT

Two-dimensional FFT (2D-FFT) is a Global Fixed algorithm. For an image $I(k,l)$, $0 <= k,l <= N$, the corresponding 2D-FFT is given by

$$F(m,n) = \sum_{k=0}^{N-1} \sum_{l=0}^{N-1} I(k,l) \, e^{-2\pi j(km+ln)/N}, \quad 0 <= m,n <= N-1$$

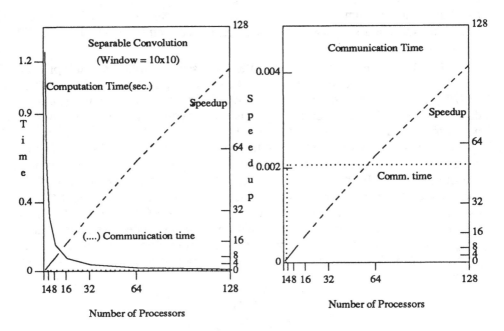

Figure 4.6 : Performance of Separable Convolution on a Processor Cluster

where j = $\sqrt{-1}$. A nice property of the 2D-FFT is that it can be performed in two decomposable steps : a one dimensional N point FFT along the rows followed by a one dimensional N point FFT of the intermediate results along the columns, or vice versa. We use this property to map 2D-FFT on the cluster processors. The algorithm consists of three phases : 1D-FFT computation along rows, transposing the intermediate results and, 1D-FFT along the columns.

Figure 4.7 describes the algorithm. In the first phase each processors is assigned N/P rows. Let's denote the sequence of rows with processor P_i as Granule(i). Also, let's divide each granule into P equal blocks of size N^2/P^2 as shown in Figure 4.8. A block B(i,j) denotes a block of size N^2/P^2 with processor P_i , $0<=j<=P-1$. Each processors computes the 1D-FFT along the rows of its granule. Then in the second phase, the processors communicate with each other in the following manner to transpose the intermediate results. A processor P_i sends block B(i,j) to processor P_j for all $0<=j<=P-1$, $j{\neq}i$. Each processor needs to communicate and exchange a block with every other processor in the cluster. However, by performing the communication systematically, the transpose can be achieved without any conflicts as described in the

---

ALGORITHM 2D-FFT

Each processor P_i receives granule(i) of rows.
/* The following description is with respect to processor P_i */

$$M := \left\lceil \frac{N^2}{P} \right\rceil$$

For k = 1 to M do
    compute 1D-FFT of row(k) of granule(i)

For j = 1 to M do (i ≠ j)
    k = i+j mod P
    connect P_i to P_k
    send Block(i,j) to P_k
    receive Block(j,i) from P_k

For k = 1 to M do
compute 1D-FFT of row(k) of granule(i)

END 2D-FFT

Figure 4.7 : 2D-FFT

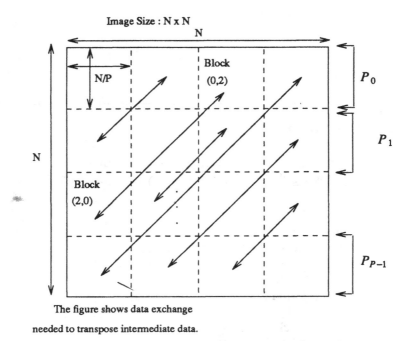

The figure shows data exchange

needed to transpose intermediate data.

Figure 4.8 : An Example of Mapping 2D-FFT onto Four Processors

algorithm. Finally, each processor computes 1D-FFT along the columns.

The 1D-FFT for size N can be done in O(NlogN) time [53]. The constant of multiplication is 6, i.e., to perform N point 1D-FFT it takes approximately 6NlogN floating point operations. Therefore, the computation time for the above algorithm is (for both row and column)

$$t_{cp} = \frac{12 \times t_{fl} \times N \times N \times \log_2 N}{P}$$

The communication time to transpose the intermediate results is

$$t_{comm} = 2 \times (P-1) \times N^2 / P^2$$

and the number of switch settings are, $t_{sw} = P-1$.

One important observation is that even though FFT is a Global Fixed algorithm, in the above mapping both the computation and communication times reduce as the number of processors increases. In other words, both computation and communication are decomposable for parallel processing. Therefore, if the communication is achieved without conflicts (as in our case), we can obtain linear speedups.

Figures 4.9 and 4.10 show the performance 2D FFT on a processor cluster. From Figure 4.9 we can observe that almost linear speedup can be obtained. The variation of the communication time as a function of the processor is shown in Figure 4.10. Note that the communication time curve follows the computation time curve in its shape and the communication is completely decomposable.

4.3.4. Hough transform

Hough transform is global varying algorithm. Also, the communication cannot be decomposed. Hough transform is a method to detect shapes such as straight lines, curves, circles, and ellipses in an input image [54]. The method is to perform the computation in the parameter space of the curves. For detecting line segments, normally the computation is done in the (r,θ) parameter space. If there exists a line whose normal distance from the origin is r, the normal makes an angle θ with the x-axis then, if the point (x,y) lies

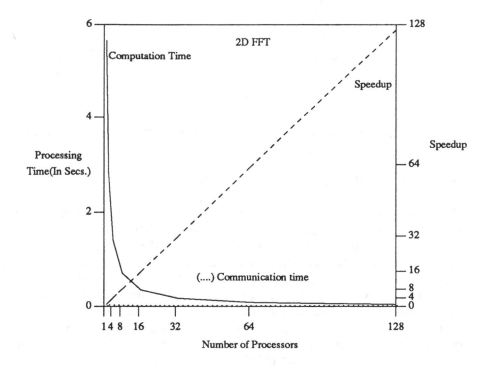

Figure 4.9 : Performance of 2D FFT on a Processor Cluster

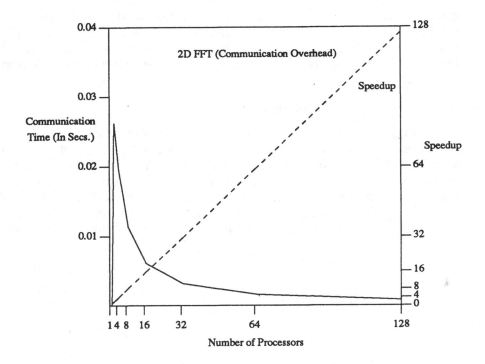

Figure 4.10 : Communication Time for 2D FFT on a Processor Cluster

on that line then the following equation is satisfied.

$$r = x\cos\theta + y\sin\theta$$

First r, θ are quantized. The quantization depends on how much accuracy is required in the final result. Assume that the maximum value of r br r_{max} maximum value of θ be θ_{max} (generally π). Then if r_{res}, θ_{res} are the resolutions used for quantization, the total number of accumulator cells in the computation are $r_{max}.\theta_{max}/r_{res}.\theta_{res}$, the number of rows and columns in the accumulator array being $\theta_c = \theta_{max}/\theta_{res}$ and $\rho_c = r_{max}/r_{res}$, respectively. The algorithm involves two major steps. The first step is to accumulate votes in the accumulator array for various digitized r and θ values. The second step is to compute local maxima in the output of the first step. The first step is regular and suitable for SIMD implementation. The second step is more suitable for MIMD implementation because the output is global data dependent. For example, an image containing many lines will result in many more maxima than an image containing a few lines, and therefore, the required computation will vary. Hence, Hough transform is a hybrid algorithm containing both SIMD and MIMD algorithms.

We present two mappings of the Hough transform algorithm for parallel processing on the processor cluster. The first mapping divides the input image into as many granules as the number of available processors. The second mapping divides the tasks depending on the parameters and desired quantization. The former is referred to as "data partitioning" and the latter as "parameter partitioning." We discuss advantages and disadvantages of both the mappings and also compare the computation time, communication time and memory requirements for both mappings.

Data Partitioning

Assume that the input image is NxN, and to simplify the discussion assume that the number of available processors is $P = p^2$. The image is partitioned into N^2/p^2 blocks. Processor $P(i,j)$ works on block $i*p + j$, where $1 \leq i, j \leq p$. Each processor computes the vote count for its part of the image for all quantizations of θ values. Figure 4.11 shows the accumulator array for a processor. Note that each processor has to maintain a complete accumulator array of size $\rho_c \times \theta_c$ and update the appropriate vote count computed from its share of the image. The algorithm ACCUMULATE_COUNT in Figure 4.12 shows the computation for this step. The computation time to compute the accumulator array is time taken to perform $2 \times \left\lceil \dfrac{n^2}{p^2} \right\rceil \times f \times \theta_c$ multiplications and half as many additions, where f is the largest fraction of significant pixels in

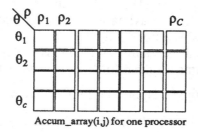

Accum_array(i,j) for one processor

Figure 4.11 : Accumulator Array for Hough Transform

ALGORITHM ACCUMULATE_COUNT
Each processor P_i, $1 \leq i \leq p^2$ does the following (in parallel)
For j = 1 to θ_c do
For each (x,y) in the subimage such that (x,y) is significant do
/*significant means black pixel or edge element*/
compute $r(\theta_j) = x \cos\theta_j + y \sin\theta_j$
Accum_array(θ_j,r(θ_j)/r_{res}) = Accum_array(θ_j,r(θ_j)/r_{res}) + 1
End_For
End_For
END ACCUMULATE_COUNT

Figure 4.12 : Algorithm to Compute Votes in Hough Transform

---

a block and θ_c is the number of quantizations for θ. The next step is to combine the partial results of all the processor to obtain a global accumulator array so that maxima can be determined. For combining the partial results we propose the tree sum method in which, at each step, twice as many processors combine their partial results, therefore requiring $2 \times \log p - 1$ steps.

The algorithm ACCUMULATE_SUM in Figure 4.13 performs the merging of partial results. The processors are numbered from 0 to $p^2 - 1$. A processor with number k, $0 \leq k \leq p^2 - 1$ corresponds to a processor (i,j) such that $k = i*p + j$.

Following this step, processor P_0 has the entire accumulator sum. The next step is to distribute this global accumulator sum to all the processors so that computation for local maxima can be performed in parallel. This step needs only one step. Processor P_0 broadcasts the entire array to all the processors using the broadcast facility of the crossbar. After the broadcast step, each processor performs a search for local maxima on its share of the accumulator rows $\left\lceil \frac{\theta_c}{p^2} \right\rceil$. In this algorithm, for each entry in its block of the accumulator array, the processor determines whether the entry represents a local maxima in a neighborhood.

In summary, the total computation and communication time requirements for the entire hough transform algorithm using the data partitioning are as follows.

/* $Accum\_array_k(i,j)$ denotes the accumulator cell (i,j)
the Accumulator array of processor k. */

ALGORITHM ACCUMULATE_SUM

For i = 0 2×log$_2$p−1 do
  For all processors P_j do in parallel ($0{\leq}j{\leq}p^2{-}1$)
    If j mod $2^{i+1} = 2^i$ then
      Connect P_j --> P_{j-2^i}
      For k = 1 to θ_c do
      For l = 1 to ρ_c do

        Send $Accum\_array_j(k,l)$ P_{j-2^i}
        $Accum\_array_{j-2^i}(k,l) := Accum\_array_{j-2^i}(k,l)$
          $+ Accum\_array_j(k,l)$
      End_For
      End_For
    End_If
  End_For
End_For
END ACCUMULATE_SUM

Figure 4.13 : Algorithm to Accumulate the Vote Count

$$t_{cp} = 3{\times}t_{fl}{\times}\left\lceil \frac{N^2}{p^2} \right\rceil {\times}f{\times}\theta_c + \theta_c{\times}C{\times}logP + \theta_c{\times}\rho_c{\times}w^2/p^2$$

where, the first term is for computing the votes, the second term is to sum the accumulator array and the third term is for looking for local maxima in a window of size w^2. The communication time for this algorithm is

$$t_{comm} = (logP+1){\times}\theta_c{\times}\rho_c$$

and the number of switch settings are $t_{sw} = logP+1$.

Unlike 2D-FFT, the communication is not decomposable. In other words, the communication increases as the number of processors increases in a cluster. In the following mapping we will observe that it is possible to reduce the communication such that instead of it increasing as a function of number of processors in the cluster, the communication remains constant.

Figure 4.14 shows the computation and communication time along with the speedup for hough transform. Even though the computation time for hough transform decreases as the number of processors increases, the computation is not completely decomposable. The second term (to combine partial results) of t_{cp} increases as a log function of the number of processors. Furthermore, the communication overhead to combine accumulator arrays also increases logarithmically with the number of processors. Consequently, for a large number of processors, the communication time becomes comparable to the computation time (as shown in Figure 4.14), and that results in degradation in speedup. We will observe in the following that it is possible to obtain almost linear speedups.

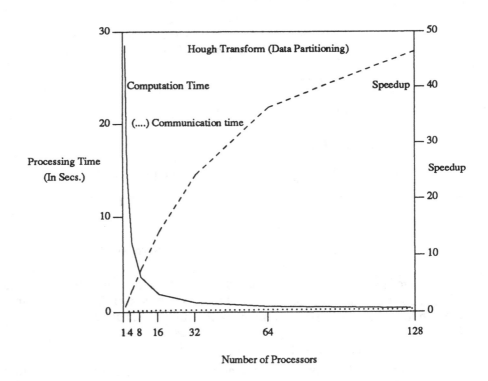

Figure 4.14 : Performance of Hough Transform (Data Partitioning)

Parameter Partitioning

In this mapping, instead of partitioning the data to the processors the parameters space is partitioned. Each processor works on the entire image but computes the vote count for only few θ values. Each processor computes all r values for its share of θ values. If there are p^2 processors, then each processor gets $n = \theta_c/p^2$ values of θ to work on. Therefore, processor i gets to work on n values of θ, where $1 \le i \le p^2$. There are several advantages to this mapping, both in terms of communication and implementation at each processor. First of all, when looking for peaks later, a processor needs to communicate with only two other processors to obtain the upper and lower boundary rows of the Accumulator array. Secondly, we introduce additional data structures to make the search for local maxima efficient, where instead of searching for the local maxima in the entire accumulator array, only a fraction indicating possible local maxima need to be searched. Furthermore, the processor can store $\sin\theta$, $\cos\theta$ values for its allocated n values of θ in its registers, since only few values need to be stored. This results in saving on local memory accesses delays which would occur if all quantized $\sin\theta$ and $\cos\theta$ values are stored with each processor in its local memory. The algorithm to compute the accumulator array at each processor is similar to that in the case of data partitioning, except that each processor works on the entire image but only its own part of the parameters.

A brief explanation of the algorithm is as follows. In the first step (computing votes), the algorithm computes value of ρ for each significant pixel for all θ values. It then increments the appropriate count in the Accumulator array. If the count increases beyond a certain threshold value, there exists a possibility of this being a local maxima. Therefore, another array called the Link_array is updated marking this fact. This step reduces the search space when looking for local maxima since normally a very small fraction of the image contributes to lines and entire Accumulator array need not be searched when looking for local maxima. Once the above computation is finished for the entire image, processor P_i communicates with P_{i+1} and P_{i-1} to obtain the boundary rows of the Accumulator array. Then the local maxima are computed in the Accumulator array using the information available in Link_array. There is a need to search only those entries in the Accumulator array for a local maxima which are marked by the Link_array. The computation, communication and memory requirements for this mapping are as follows.

$$t_{cp} = 3 \times t_{fl} \times \left\lceil \frac{N^2}{p^2} + 1 \right\rceil \times f \times \theta_c + \theta_c \times \rho_c \times w^2/p^2$$

where the first term is for computing the votes and the second term is to for local maxima in a window of size w^2. The communication time for this algorithm is

$$t_{comm} = 2{\times}\rho_c$$

and the number of switch settings are $t_{sw} = 2$.

The memory requirements of the two partitionings are comparable. For example, for a typical image size of $512x\,512$, value of ρ_c will typically be $512{\times}\sqrt{2}$, and C will be 180. However, each pixel normally is a byte where as each accumulator cell is an integer. Assuming a 4 byte integer, in data partitioning a processor has to store the entire accumulator array of size 521 Kbytes (approximately), and in the second mapping a processor has to store the entire image (256 K bytes) and its part of the accumulator array.

There is another way in which the parameter partitioning mapping can be performed. Instead of storing the image in all the processors, a controller processor, such as a DSP, can store the image and broadcast each significant pixel value and its location while processors compute the votes in an SIMD lock-step fashion. This results in saving the memory, because now only one processor need store the image. We make the following observations. The communication requirement is $f{\times}N^2$, where f is the fraction of significant pixels. However, the communication can be overlapped with computation because while processors are computing the vote count for a location in the image, the next location can be broadcast. Therefore, the time to compute the Accumulator_array in this case will be MAX(t_{cp} , Broadcast time for $f{\times}N^2$ pixels locations).

By using parameter partitioning the overhead of combining partial results is eliminated, and for each processor the communication is reduced to exchanging one row of the accumulator array with two other processors. Therefore, the communication remains constant as the number of processors increases. Figure 4.15 shows the speedup, computation time and communication time for hough transform using parameter partitioning. Figure 4.16 compares the communication overhead and the speedup for the two types of partitioning. Notice that using parameter partitioning it is possible to obtain almost linear speedup.

4.4. Parallel Implementation Results

This section contains implementation of some algorithms on a simulated processor cluster. A cluster was simulated on an intel iPSC/2 hypercube multiprocessor. The performance results capture all the overheads associated with parallel programming, and therefore, the results are very accurate. Also, we show through the example of 2-D FFT algorithm that the analysis

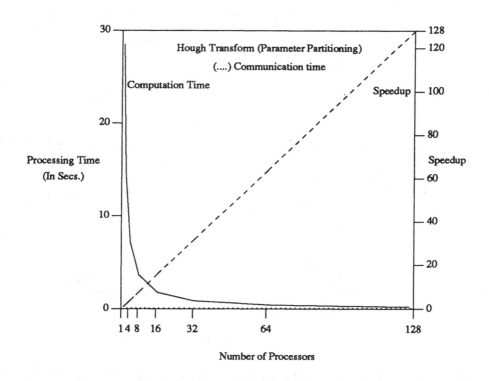

Figure 4.15 : Performance of Hough Transform (Param. Partitioning)

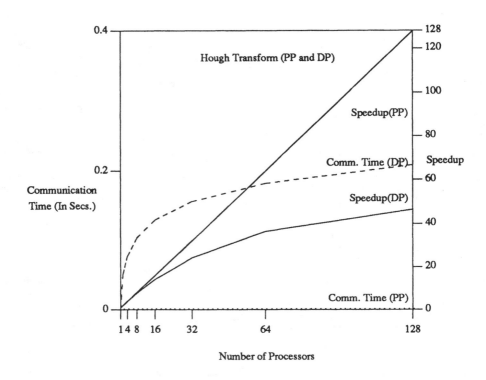

Figure 4.16 : Comparison of Performance of PP and DP for Hough Transform

presented in the previous section is very close to the implementation results. We present performance results for four algorithms in this section. Two algorithms are 2-D FFT and separable convolution. The other two algorithms are parts of the Image Understanding Benchmark Algorithms developed by Weems et al [1]. The two algorithms are sobel edge detection and median filtering. The performance of the algorithms has been evaluated using the test data provided with the benchmark algorithms [1].

4.4.1. 2-D FFT

A mapping of 2-D FFT has been described in Section 4.1. Figure 4.17 shows the performance of 2-D FFT on a 16 processor cluster (image size 256x256). Other parameters are the same as given in Table 4.1. Solid lines in the graph show the computation times for analysis (symbol +) and implementation. We observe that the analytical results are very accurate. However, the implementation times are a little more than that given by analysis because implementation captures the overhead of index management, etc. which is not included in the analysis. The graph also shows the corresponding speedups for both cases. Note that speedups obtained through analysis and implementation are almost the same and are practically indistinguishable. Figure 4.18 shows graphs for the communication time. Again, implementation and analytical results are very close to each other.

4.4.2. Separable convolution

Table 4.2 shows the performance for separable convolution implementation on a 256×256 image with window size 10×10. The table shows the major computation operations in the algorithm which include floating point operations as well as integer operations. The fifth column shows the number of times connection in the crossbar needs to be changed during the algorithm

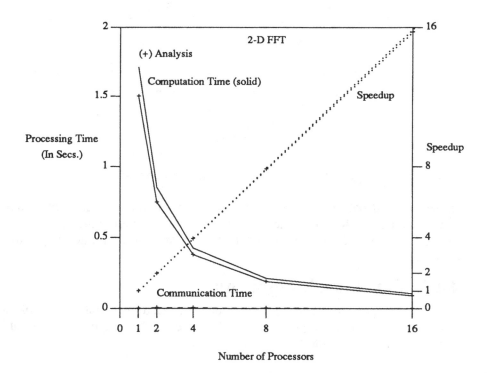

Figure 4.17 : Performance of 2D-FFT (Analysis and Implementation)

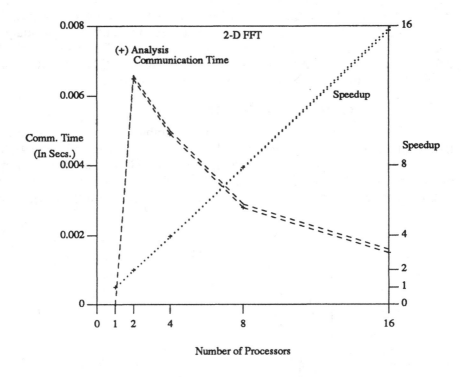

Figure 4.18 : Communication Time for 2-D FFT

execution, and column 6 contains the rounded value of the amount of data communicated in KBytes. The table shows that the communication time is very small compared to the computation time, and therefore, good speedups are obtained.

4.4.3. Benchmark Algorithms

The Image Understanding Benchmark provided the serial version of the programs and the data [1]. We implemented sobel edge detection and median filtering algorithms.

Sobel

Sobel edge detection is a two-dimensional convolution operation with a 3×3 mask. The implementation was done using medium grain parallelism in

Table 4.2 : Separable Convolution Implementation Results

Separable Convolution						
Window 10x10						
No. Proc.	Fl. Point K. Ops	Other K. Ops	Comp. Time (ms.)	Comm. Setup	Comm. K Bytes	Comm. Time(ms.)
1	3932	3932	2607	0	0	0
2	1966	1966	1310	2	20	4.09
4	983	983	658	3	20	4.09
8	492	492	332	3	20	4.09
16	246	246	169	3	20	4.09

an MIMD mode, and mapping was similar to that of separable convolution. Table 4.3 illustrates the performance results for sobel edge detection algorithm. There were six data sets but here we present results using only one data set (test, size 256×256). The results obtained on other data sets were similar. The table includes all overheads, including program load time, data load time, data input time (from global memory), and time to gather results. If all the overhead is included, then the performance for larger cluster size is sublinear. There are two main reasons for this performance. First, amount of computation per pixel is very small (3×3 convolution), and secondly, all the overhead is included in the computation of the speedup. The parameters for communication bandwidth are conservative (20 MBytes/sec.) and if the bandwidth is assumed to be larger, then the performance is expected to be much better.

Median filtering

Table 4.4 shows the performance results for the median filtering algorithm. The algorithm was evaluated on the same data set. Size of the median filter was 5×5. Data is partitioned along the rows. Each processor is allocated an equal number of rows and two boundary rows in each direction. There is no need for communication during the algorithm execution. Median filtering does not involve any floating point multiplication or addition operations (only comparison operations are needed). Table 4.4 shows that we can obtain good speedups on a cluster for median filtering.

Table 4.3 : Sobel Edge Detection

Sobel (Test)							
No. Proc.	Proc. Time(sec.)	Data load Time(Sec.)	Result Output Time(sec.)	Prog. Load Time(sec.)	Data Input Time(sec.)	Total Time(sec.)	Speed Up
1	4.04	0	0	0	0.008	4.05	1
2	2.02	0.056	0.014	0.001	0.008	2.1	1.92
4	1.01	0.056	0.014	0.001	0.008	1.09	3.70
8	0.51	0.056	0.014	0.001	0.008	0.589	6.91
16	0.26	0.056	0.014	0.001	0.008	0.33	12.13
32	0.13	0.056	0.014	0.001	0.008	0.21	19.71

Table 4.4 : Median Filtering

Median Filtering (Test)							
No. Proc.	Proc. Time(sec.)	Data load Time(Sec.)	Result Output Time(sec.)	Prog. Load Time(sec.)	Data Input Time(sec.)	Total Time(sec.)	Speed Up
1	60.36	0	0	0	0.008	60.37	1
2	30.17	0.056	0.056	0.001	0.008	30.30	1.99
4	15.19	0.056	0.056	0.001	0.008	15.31	3.94
8	7.72	0.056	0.056	0.001	0.008	7.85	7.70
16	3.99	0.056	0.056	0.001	0.008	4.11	14.68
32	1.90	0.056	0.056	0.001	0.008	2.02	29.93

4.5. Summary

To evaluate parallel algorithms on a cluster, we explored alternative mapping strategies and computation modes. Some of the algorithms have been implemented on a simulated cluster, and we show that the analysis provides very accurate results. The performance results show that very good speedups can be obtained on a processor cluster in any computation mode. The parameters chosen for processor speed and communication speed were very conservative. We think that much faster processors and communication links are possible and available with today's technology, and therefore, the performance results presented in this chapter are also conservative.

Chapter 5

Inter-Cluster Communication In NETRA

The focus of this chapter is inter-cluster communication in NETRA and performance evaluation of parallel algorithms when mapped across multiple clusters. When an algorithm is mapped on multiple clusters, processors belonging to different clusters may need to communicate, and therefore, inter-cluster communication is needed. However, unlike intra-cluster communication between processors, there may be conflicts in accessing the global interconnection network, global memory or other common resources. These conflicts need to be taken into account when computing the inter-cluster communication, and consequently, performance of algorithms when mapped across multiple clusters is affected by conflicts. In this chapter we present a method to evaluate inter-cluster communication time under the presence of conflicts. The method is based on the work by Patel [55,56].

This chapter is organized as follows. Section 5.1 presents alternative inter-cluster communication strategies in NETRA. Analysis of inter-cluster communication strategies is presented in Section 5.2. How the analysis can be incorporated into the performance evaluation of algorithms is the subject of Section 5.3. Section 5.4 contains the performance evaluation of various algorithms whose performance on one cluster was presented in the preceding chapter. Finally, Section 5.5 summarizes the chapter.

5.1. Alternatives for Inter-cluster Communication

5.1.1. Multistage interconnection network and global memory

In this method global memory is used for inter-cluster communication. The global memory is accessed through the multistage interconnection network by processors in a cluster or by a DSP. A processor(s) needing to send data to another processor in a different cluster writes the data into designated locations in the memory. This involves setting the appropriate circuit through the global multistage interconnection network to the memory module followed by a data transfer. The data is transferred in block mode. The Memory Line Controller (MLC) updates the information about the destination port(s), length of the data block, and block's starting address, and sets a flag indicating the availability of data. Now the destination processor can read the data using this information. Note that this method permits out of order requests to be serviced. For example, if the destination processor tries to read the data before it has been written, the MLC informs the processor of this situation, and when the data is really written into the global memory then the MLC informs the destination processor. This is a block level data-flow approach. The main advantages of this approach are that asynchronous communication is possible, out of order messages can be handled, and efficient pipelining of data can be achieved. This method is depicted in Figure 5.1. The Figure shows how a processor P_i of cluster C_i will communicate with processor P_j of cluster C_j using the strategy.

5.1.2. DSP tree links

The second alternative to performing inter-cluster communication is to use the DSP tree links. However, for distant inter-cluster communications, the tree may not perform well because of the root bottlenecks typical to any tree structure. The main function of the tree structure is to provide control hierarchy for the clusters. Its links are mainly used for program and data broadcast to subtrees, and DSPs use the tree links to send (receive) control

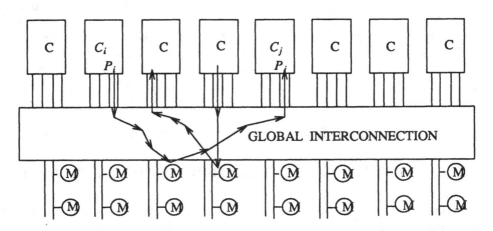

Figure 5.1 : Inter-cluster Communication Using Global Memory

information to (from) other DSPs. Therefore, the DSP tree is designated mainly for control function, and we do not envision it to be used for large data transfers between distant processors.

5.1.3. Global bus

The third alternative strategy to perform inter-cluster communication is to use a high speed global bus that connects all DSPs and one port from each cluster. The global memory is also connected to the bus and is accessible to all clusters via the bus. Note that the global bus is proposed to be an alternative global interconnection to the multistage interconnection network. If the bus can be designed fast enough (such as by using fiber optics), and if inter-cluster communication is low, the global bus presents a viable alternative to the multistage interconnection network. In this scheme, the communication is done explicitly by messages, and it is synchronous. Figure 5.2 show this communication method. The Figure shows how a processor P_i of cluster C_i will communicate with processor P_j of cluster C_j using the strategy.

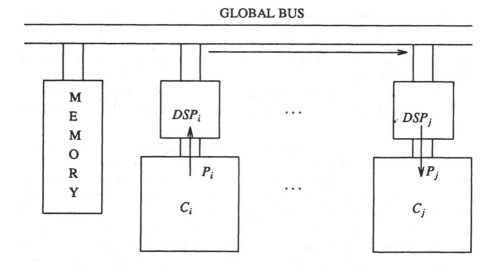

Figure 5.2 : Inter-Cluster Communication Using the Global Bus

5.2. Analysis of Inter-cluster Communication

Inter-cluster communication is needed in the following cases : i) An algorithm is mapped in parallel on more than one cluster and the processors need to communicate to exchange partial results or combine their results. ii) In an integrated vision system, output data of a task produced at one or more clusters needs to be transferred to the next task executing on different clusters. iii) It is needed to perform input and output of data and results. The amount of inter-cluster communication depends on the type of algorithms, how they are mapped in parallel, frequency of communication and amount of data to be communicated.

There are several parameters that affect the inter-cluster communication time. The architecture dependent parameters are the number of processors (i.e., number of clusters and number of processors in each cluster), number of memory modules, number of processors per port connected to the global interconnection, and the type of interconnection network. Some parameters depend both on the architecture as well as on the type of algorithms, how they are mapped, and their communication requirements when mapped onto multiple clusters. Furthermore, not only does the communication time depend on the underlying algorithms but it also depends on the network traffic generated by other processors in the system because there may be conflicts in accessing the network as well as memory modules.

We consider an equivalent model of the architecture as shown in Figure 5.3. The model shows N processors connected to M memory modules through a global interconnection network. N is given by $C \times P_t + N_{dsp}$, where C is the number of clusters, P_t is the number of ports in each cluster and N_{dsp} is the number of DSPs in the system. For simplicity, we assume that each cluster contains an equal number of processors. The number of physical processors will be given by $C \times P_t \times P_p$, where P_p is the number of processors per port.

The following analysis is based on the analysis presented by Patel in [55, 56]. He developed an analytical model for evaluating alternative processor memory interconnection performance and showed that the analysis is reasonably accurate. Consider execution of a typical parallel algorithm on multiple clusters. The execution will consist of processing, intra-cluster and inter-cluster communication. Figure 5.4 shows the computation and communication phases of an algorithm. The computation time is given by t_{cp}, the intra-cluster communication time is given by t_{cl}, and the inter-cluster communication time is given by t_{icl} in terms of equivalent processor cycles. However, due to conflicts in the network or conflicts in accessing memory modules, a processor may have to wait for w_a cycles before being able to

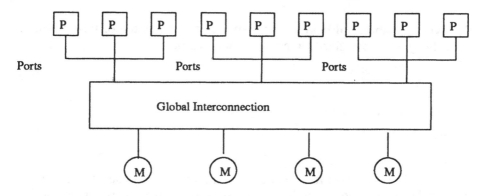

Figure 5.3 : Equivalent Model for Global Communication

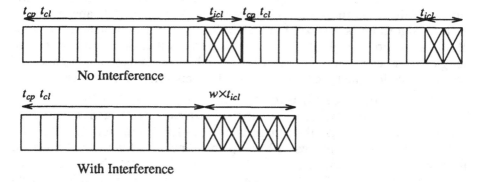

Figure 5.4 : Computation and Communication Activities

access the network and write to (or read from) the memory. In effect, this can be seen as the communication time being elongated by a factor w for each request, and instead of communication time being t_{icl}, it is now $w \times t_{icl}$ as shown in Figure 5.4. Therefore, if the probability of accessing the global network in each processing cycle is m and for each access the communication time is t_{icl}, then the useful computation for t processor cycles takes $t + m \times t \times w \times t_{icl}$, where $t = t_{cp} + t_{cl}$. The fraction of useful work (utilization U) is given by

$$U = \frac{t}{t + m \times t \times w \times t_{icl}}. \tag{5.1}$$

The average number of busy memory modules (or fraction of time when the bus is busy when the global interconnection is a bus) is

$$B = \frac{N \times m \times t_{icl} \times t}{t + m \times t_{icl} \times t \times w} \tag{5.2}$$

and in terms of utilization,

$$B = N \times m \times t_{icl} \times U. \tag{5.3}$$

In [55], it is shown that the utilization primarily depends on the product $m \times t_{icl}$ rather than m and t_{icl} individually. In other words, the processor utilization primarily depends on the traffic intensity and to a lesser extent on the nature of the traffic.

For a particular algorithm, all the parameters are known except w. The probability of accessing the global network is essentially given by the number of times communication is needed per processor cycle and is known when an algorithm is mapped in parallel. The factor w depends on the algorithm parameters as well as the interference from other processors accessing the global network and the memory, number of processors, number of memory modules, the type of interconnection network and the access rate m.

Consider the processor activities again. A processor needing to access the global memory or the bus submits requests again and again until accepted; on an average this happens for $(w-1) \times t_{icl}$ time units. After the request is granted, the processor has a path to memory for t_{icl} time units. In other words, the network sees an average of $w \times t_{icl}$ consecutive requests for unit service time. Therefore, the request rate (for unit service) from a processor as seen by the network is

$$m' = \frac{m \times w \times t_{icl}}{1 + m \times w \times t_{icl}}. \tag{5.4}$$

and in terms of utilization
$m' = 1 - U$.
For details, the reader is referred to [55].

The model that we analyze is a system of N sources and M destinations. Each source generates a request with probability m' in each unit time. The request is independent, random, and uniformly distributed over all destinations. Each request is for one unit service time. The following is an analysis for a bus and for multistage delta network.

Bus : We know from earlier discussion (Equation 5.3) that

$$B = N \times m \times t_{icl} \times U \tag{5.5}$$

and also, assuming all sources have the same request rate, average amount of time the bus is busy is given by

$$B = [1 - (1 - m')^N]. \tag{5.6}$$

Equations 5.4 and 5.5 result in a non-linear equation

$$N \times m \times t_{icl} \times U - [1 - (1 - m')^N = 0. \tag{5.7}$$

In the above equation, value of m' can be substituted in terms of w, and hence, value of w can be computed. If the request rate from sources is not uniform, i.e., if the request rate from source N_i is m_i then the above equation becomes

$$\sum_{j=1}^{j=N} (m_i \times t_{icl}(j) \times U_j) - [1 - \prod_{j=1}^{j=N}(1 - m'_j)] = 0. \tag{5.8}$$

When evaluating performance of a parallel algorithm mapped across clusters there will be two request rates, one for the processors taking part in executing the algorithm and the other for the rest of the processors in the system which will be an input parameter.

Multistage-Interconnection (Delta) : A delta network is an n stage network constructed from $a \times b$ crossbar switches with a resulting size of $a^n \times b^n$. Therefore, $N = a^n$ and $M = b^n$. For a complete description refer to [56]. Functionally, a delta network is an interconnection network which allows any of N sources (processors) to communicate with any one of the M destinations (memory modules). However, two requests may collide in the network even if the requests are made to different memory modules. We use results from [55, 56] to obtain the average number of busy main memory modules B, which is given by

$$B = M \times m_n \tag{5.9}$$

and the following equation in satisfied.

$$N \times m \times t_{icl} \times U - M \times m_n = 0 \tag{5.10}$$

where,

$$m_{i+1} = 1 - (1 - \frac{m_i}{b})^a, \ 0 \leq i < n$$
and, $m_0 = 1 - U.$

For details, the reader is referred to [55, 56].

These equations are solved numerically to obtain the interference delay factor w which is used in the performance evaluation of algorithms mapped across multiple clusters.

5.3. Approach to Performance Evaluation

Performance of an algorithm mapped on multiple clusters is governed by various factors. Table 5.1 summarizes the parameters affecting the performance of a parallel algorithm. The approach to evaluating the performance of an algorithm is as follows. Using the parameters and a particular mapping, computation (t_{cp}), intra-cluster communication (t_{cl}) and inter-cluster communication time (t_{icl}) are determined. The traffic intensity for a processor(s) (or a cluster depending on how an algorithm is mapped) is given by $\dfrac{t_{icl}}{t_{cp}+t_{cl}}$. Using the traffic intensity values, and using a range of traffic intensity values for interference, the effective bandwidth of the network is determined; that is, the factor w is computed.

Consider a parallel execution of an algorithm across clusters. If the execution time when the algorithm is executed on a single processor is t_{seq}, then the speedup in the best case is given by

Table 5.1 : Parameters for Performance Evaluation

Total No. of Processors N_p	512
Cluster Size P_c	8-128
No. of Processors/Port P_p	4
Image Size $N{\times}N$	512 X 512
Memory Modules M	128
Processor Speed	5 MIPS, 5 MFLOPS
Network Speed (Block Transfer)	20 Mbytes/Sec.
Traffic Intensity for Interference $(m{\times}t)$	0.1,0.4,0.8

$$Sp = \frac{t_{seq}}{t_{cp} + t_{cl} + t_{icl}}, \tag{5.11}$$

that is, assuming there is no interference while accessing the network or the global memory. Under the conditions in which there are conflicts while accessing the network, the inter-cluster communication time is given by $w \times t_{icl}$, and therefore, the speedup is given by

$$Sp' = \frac{t_{seq}}{t_{cp} + t_{cl} + w \times t_{icl}}. \tag{5.12}$$

Hence, degradation in speedup with respect to the best case speedup will be

$$\frac{Sp - Sp'}{Sp} = \frac{(w-1) \times t_{icl}}{t_{cp} + t_{cl} + w \times t_{icl}}. \tag{5.13}$$

5.4. Performance of Parallel Algorithms on Multiple Clusters

The extent of inter-cluster communication depends on the type of algorithms, how they are mapped in parallel, frequency of communication, and amount of data to be communicated. As discussed in the previous chapter, these requirements vary for algorithms belonging to different classes.

We are mainly interested in the performance evaluation of parallel algorithms when mapped across clusters. The performance of an algorithm will be affected by interference from other processors in the system which are not executing the particular algorithm under study.

This section discusses the performance of various algorithms when mapped across clusters. The algorithms are selected according to their communication requirements. We have chosen one algorithm from each of the following categories: Local Varying, Global Fixed and Global Varying. Algorithms in each of these categories exhibit different communication characteristics, and therefore, the analysis will provide the performance of the architecture for a wide range of algorithms.

5.4.1. Two-dimensional Fast Fourier Transform (2-D FFT)

From Chapter 4 we know that a 2-D FFT can be performed in two steps : a one-dimensional N point FFT along the rows followed by a one-dimensional N point FFT of the intermediate results along the columns, or vice versa. We use this property to map the algorithm across clusters. Hence, dividing the data along rows will not require communication when computing one-dimensional FFT. However, communication is needed to obtain transpose of the intermediate results. Figure 5.5 shows an example of the two steps and communication for three clusters.

Clusters are allocated rows in proportion to their size. A cluster C_i of size $P_c(i)$ (i.e., containing $P_c(i)$ processors) is allocated $\dfrac{N \times P_c(i)}{\sum_{i=1}^{i=n} P_c(i)}$, where n is the total number of clusters executing the algorithm. Within a cluster rows are equally divided among processors. In the first phase processors compute N point FFT of all the rows in their granule. In the second phase, to obtain transpose of the intermediate data, processors write the intermediate results into the designated global memory locations, which is read by other processors. Data remaining within a cluster is transposed using the cluster crossbar.

The computation time in terms of number of instructions is given by the following. The total number of processors are given by P, and we assume all clusters have the same size (P_c).

$$t_{cp} = \frac{12 \times N^2 \log_2(N) \times t_{fl}}{P} \tag{5.14}$$

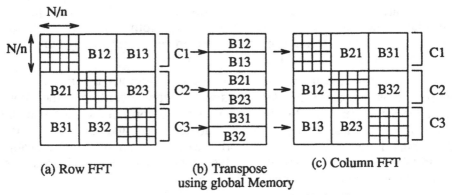

(a) Row FFT   (b) Transpose   (c) Column FFT
using global Memory

The shaded area denotes data which remains within the cluster

Figure 5.5 : An Example of Mapping 2-D FFT on Three Clusters

where, t_{fl} is the number of instructions per floating point operation. The intra-cluster communication time (t_{cl}) and the inter-cluster communication time (t_{icl}) are given by

$$t_{cl} = \frac{2 \times N^2 (P_c - 1)}{P^2} \qquad (5.15)$$

$$t_{icl} = \frac{4 \times N^2 \times (n-1) \times P_p \times R}{n \times p} \qquad (5.16)$$

where P_p is the number of processors per port and R is the communication speed of the network in terms of number of instructions per word transfer.

Using these parameters for 2-D FFT traffic intensity, computation times and parameters from Table 5.1, we evaluate the performance using the analysis presented earlier. Figure 5.6 shows the speedup obtained for the 2-D FFT algorithm. The X-axis shows the number of processors (cluster size is 16). For example, value 48 means that the algorithm is executed on 3 clusters, each containing 16 processors. The four different graphs in the Figure show speedups for no conflict (best case), low conflict, medium conflict and high conflict cases through the global interconnection network (multistage interconnection). Similar results are presented later in this section showing when a bus is used as the global interconnection network. It is observed that speedup obtained under varying degrees of conflicts through the network is comparable to that obtained in the best case. However, the best case speedup itself is not linear because of the delays through the network and the global memory.

Figure 5.7 shows the computation and communication time for 2-D FFT as a function of number of processors. Figure 5.8 shows a blown-up graph for the communication times. As we observe, the communication time is much smaller than the computation time. Furthermore, the communication time also decreases as the number of processors (clusters) increases. Also note that the intra-cluster communication time is much smaller than the inter-cluster communication time. Figure 5.9 shows percentage degradation in speedup, as defined in Equation (5.13), for different levels of conflict in the network. The degradation in the speedup levels off after increasing initially because the communication time decreases as the number of processors increases.

Figure 5.10 shows the sensitivity of the speedup to the network bandwidth. The network bandwidth is normalized to computation speed. For

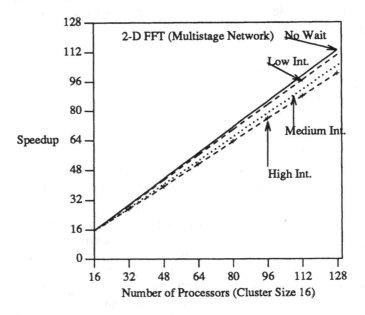

Figure 5.6 : Speedup for 2-D FFT (Multistage Network)

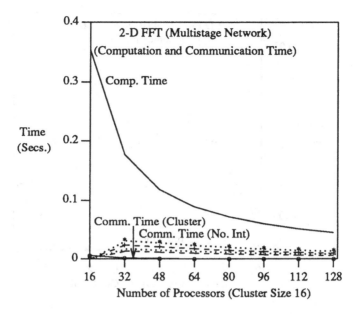

Figure 5.7 : Computation and Communication Times for 2-D FFT
(Multistage Network)

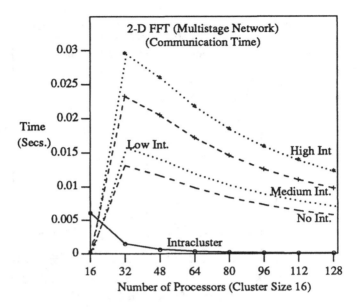

Figure 5.8 : Communication Times for 2-D FFT

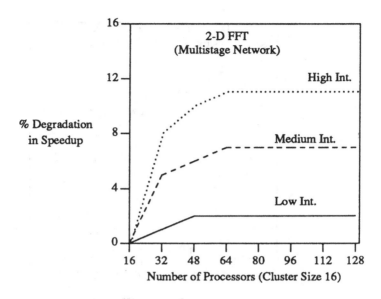

Figure 5.9 : Degradation in Speedup Due to Conflicts for 2-D FFT
(Multistage Network)

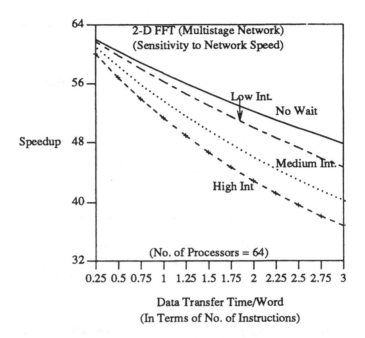

Figure 5.10 : Speedup vs. Network Speed

example, value 1 on the X-axis means that it takes the same amount of time (amortized or block in block transfer mode) to write/read a word to/from global memory as it takes to execute one instruction. The region on the left of 1 indicates faster communication network and to the right of 1 indicates slower communication network. It is evident from the Figure that degradation in speedup occurs very fast as the communication becomes slower. Therefore, in order to obtain any significant speedups from parallel computation, it is important to have matched computation and communication speeds; otherwise, increasing the number of processors or the processor speeds will not improve the performance as expected. The Figure also illustrates that the four graphs in the Figure diverge as the communication becomes slower meaning that slower performance under heavy traffic suffers more in the slower network than under light traffic.

The following is a discussion of the performance of 2D-FFT when a bus is used as a global interconnection network. The algorithm is mapped as described above. Since global bus can be accessed by only one processors at a time, the inter-cluster communication time becomes additive as the number of clusters is increased. Therefore, the performance is expected to be worse than that in the case of the multistage interconnection network. The total computation time remains the same as in the previous case and is given by

$$t_{cp} = \frac{12 \times N^2 \log_2(N) \times t_{fl}}{P}.$$
(5.17)

However, the inter-cluster communication time becomes

$$t_{icl} = \sum_{i=1}^{i=n-1} \frac{2 \times R \times N^2}{n^2} = \frac{2 \times R \times (n-1) \times N^2}{n^2}.$$
(5.18)

In other words, each cluster needs to send $\frac{(n-1)}{n}$ fraction of its data to transpose the intermediate results. This is achieved by a designated processor in each cluster, which collects the data and broadcasts it on the bus to be read by other cluster processors. Hence, there is an additional overhead of collecting and distributing the intermediate data. The intra-cluster communication time in this case is given by

$$t_{cl} = t_{cl1} + t_{cl2} + t_{cl3}$$

where,

for within cluster transpose, $t_{cl1} = \dfrac{2 \times (P_c - 1) \times N^2}{P_c^2 \times n}$, and,

for sending, receiving and redistributing the intermediate data, $t_{cl2} = t_{cl3} = \dfrac{2 \times N^2 \times (n-1)}{n^2}$.

Using these parameters, we evaluate the performance of 2-D FFT under varying degrees of conflicts on the bus. Figure 5.11 shows the speedup for 2-D FFT as a function of the number of processors (cluster size 16). When there is no conflict on the bus, the speedup increases with the number of processors. However, under conflicts, the speedup first decreases and then increases slowly. In fact, for medium and high conflicts, the speedup obtained on one cluster is better than that obtained using multiple clusters. the reason for such poor performance is that even though the communication is decomposable in 2-D FFT, the inter-cluster becomes communication time additive due to the bus and increases as the number of clusters executing the algorithm increases as shown in Figure 5.12. It is evident from the Figure that the computation time decreases but the communication time increases and becomes more than the computation time.

Figure 5.13 shows the relative performance degradation in the speedup. The degradation is very significant. However, the degradation itself decreases as the number of processors (clusters) increases because more clusters

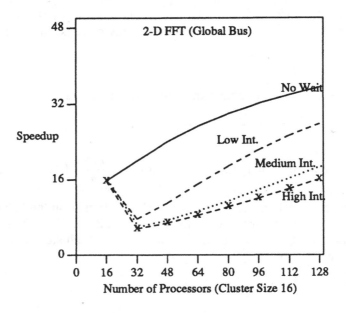

Figure 5.11 : Speedup for 2-D FFT (Global Bus)

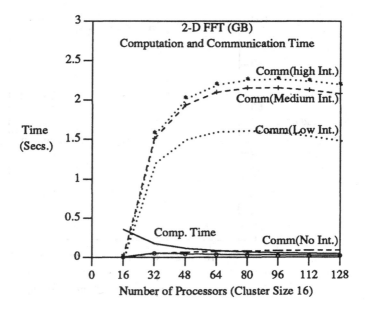

Figure 5.12 : Computation and Communication Times for 2-D FFT
        (Global Bus)

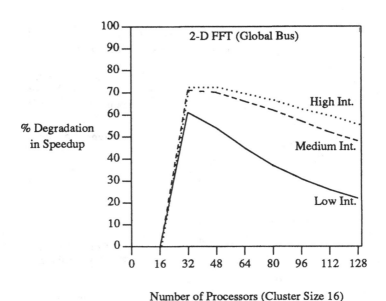

Number of Processors (Cluster Size 16)

Figure 5.13 : Degradation in Speedup Due to Conflicts 2-D FFT
(Global Bus)

execute the algorithm, and consequently, less number of clusters interfere.
Figure 5.14 shows the sensitivity of the speedup to the bus speed. Again, the
Figure shows that performance degrades rapidly as the bus becomes slower.
In order for a bus to be viable global interconnection network it is essential
that the bus bandwidth be much greater than the processor speed.

5.4.2. 2-D separable convolution

This algorithm consists of two steps. First convolution along rows
using two one-dimensional masks and then convolution along columns of the
intermediate results. Partitioning along rows in clusters, therefore, avoids
communication in the first step. However, before the second step can be performed, boundary rows with each cluster need to be communicated to other
clusters. Figure 5.15 shows the mapping on three clusters. Note that unlike in
2-D FFT, a cluster needs to communicate with at most two other clusters to
obtain the upper and lower boundary rows of the intermediate results. The
number of rows to be exchanged depends on the kernel size. For a kernel
size of $w{\times}w$, the number of rows to be exchanged along each direction is $\frac{w}{2}$.

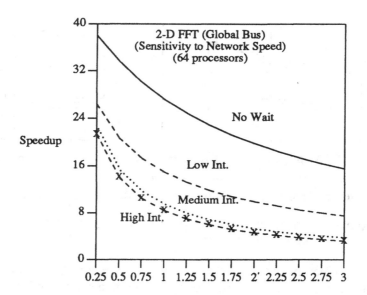

Data Transfer Time/Word
In Terms of No. of Instructions

Figure 5.14 : Speedup vs. Network Speed (Global Bus)

The amount of communication is fixed and is independent of the number of clusters on which the algorithm is mapped. The same mapping will work for regular 2-D convolution except that the amount of computation per pixel will be larger.

The computation time for the two steps is given by

$$t_{cp1} = t_{cp2} = \frac{2 \times t_{fl} \times N \times (\frac{w}{2}+1)}{P} \tag{5.19}$$

the intra-cluster communication is given by

$$t_{cl} = 2 \times N \times w, \tag{5.20}$$

and the inter-cluster communication is given by

$$t_{icl} = 2 \times w \times N \times R. \tag{5.21}$$

Figure 5.16 depicts the speedup obtained for the 2-D convolution algorithm as a function of the number of clusters (cluster size = 16). The speedup

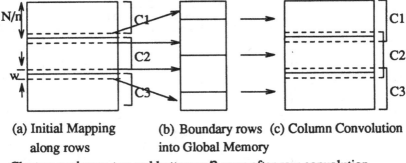

(a) Initial Mapping   (b) Boundary rows (c) Column Convolution
    along rows   into Global Memory

Clusters exchange top and bottom w/2 rows after row convolution

Figure 5.15 : An Example of Mapping 2-D Separable Convolution on Three Clusters

increases sublinearly as the number of clusters increases. The reason for not obtaining better speedup is that the computation per point of the input is small, the computation per processor decreases as the number of clusters increases, but the communication remains constant (as long as the granularity per processor is at least $\frac{w}{2}$ rows). Hence, the ratio of computation and communication decreases as the number of processors increases. The computation and communication times are shown in Figures 5.17 (a) and (b). Figure 5.16 compares the two times whereas Figure 5.17 shows only the communication time.

Note that inter-cluster communication can be avoided completely if clusters are assigned overlapped rows to perform the first step. That is, if a cluster is responsible to compute 2-D convolution for R_i rows, then its is assigned $w + R_i$ rows. Therefore, each cluster has to perform additional computation to obtain 1-D convolution of w additional rows. If the extra computation time is less than the communication time then overlapped data partitioning is better.

Figure 5.18 shows a performance comparison of the two partitioning methods. When the number of processors executing an algorithm is small, the performance is almost the same. For smaller window sizes the difference is marginal and becomes apparent only when the number of processors becomes large. However, as the window size increases (40x40 in Figure 5.18), the performance with overlapped computation becomes poor because the overhead of extra computation becomes larger than the communication

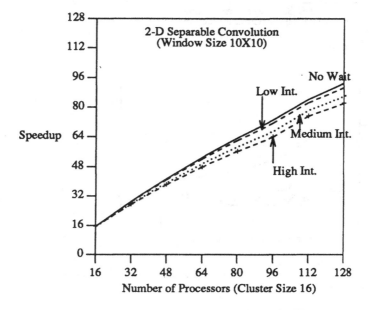

Figure 5.16 : Speedup for 2-D Convolution (Multistage Network)

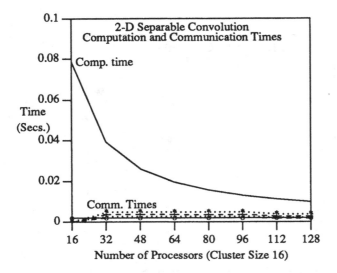

(a) Computation and Communication Time

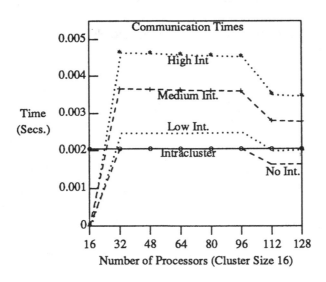

(b) Communication Times
Figure 5.17 : 2-D Convolution (Multistage Network)

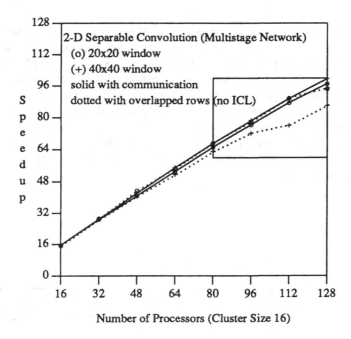

Number of Processors (Cluster Size 16)

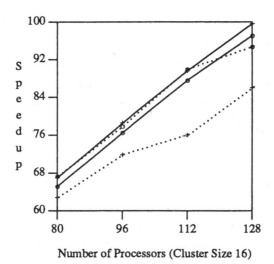

Number of Processors (Cluster Size 16)

The box in the upper graph has been blown up in the bottom graph

Figure 5.18 : Overlapped Computation vs. Communication Trade-off
(2-D Separable Convolution)

overhead.

Figure 5.19 shows the performance of the algorithm when the bus is used as a global interconnection network. The speedup increases as the number of clusters increases but eventually levels off. Though inter-cluster communication time per cluster is constant, total communication time increases as the number of clusters increases, because only one cluster can send data on the bus at any time. This is illustrated in Figure 5.20 where the communication time (with no interference) is a linear function of the number of clusters. Another reason for speedup to level off is that for a larger number of clusters the computation time becomes comparable or smaller than the communication time.

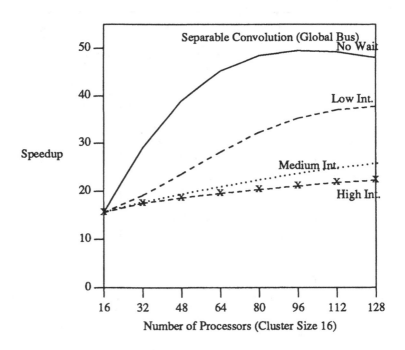

Figure 5.19 : Speedup for 2-D Convolution (Global Bus)

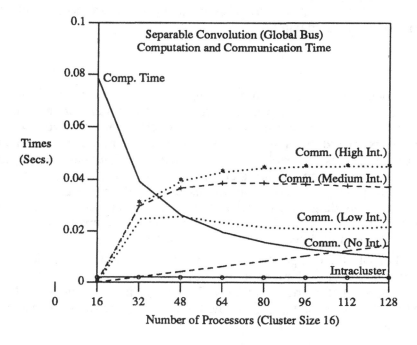

Figure 5.20 : Computation and Communication Times for 2-D Convolution (Global Bus)

5.4.3. Hough transform

We have evaluated two mappings for hough transform, namely, Data Partitioning (DP) and Parameter Partitioning (PP). The difference between the two mappings is described in Chapter 4. Briefly, in DP, data is decomposed among clusters and, in PP, parameters are decomposed across clusters.

Data Partitioning

Data is allocated to clusters in proportion to their size. Within a cluster data is distributed equally among the processors. The algorithm consists of three phases. In the first phase, each processor computes and accumulates the count contributed by its data for all the parameter values. Note that each processor maintains the entire accumulator array. In the second phase, partial results are combined within a cluster, i.e., all the accumulator arrays are added together, and then a designated processor from each cluster writes the accumulator array to designated memory locations. Arrays from all the clusters participating in the algorithm execution are then collected by one cluster. In the third phase, the cluster having the entire accumulator array computes the local maxima.

Parameter Partitioning

Under this scheme, each cluster is assigned the entire input data but is assigned only a part of the parameter space. The parameter space is partitioned in proportion to the cluster size. Each cluster receives two more parameters (boundary values) so that inter-cluster communication is avoided. That is, each cluster performs a fixed amount of additional computation to avoid communication. Within a cluster, however, data is distributed equally among the processors, and all processors work on the entire allocated parameter space. Dividing the parameter space results in mutually exclusive accumulator arrays with processors, and therefore, to compute local maxima, there is no need for inter-cluster communication.

For DP, the computation and communication times for various phases are as follows: t_{cp1} is for computing accumulator count, t_{cp2} is for combining partial accumulator arrays within a cluster, t_{cp3} is for computing the final accumulator array, and t_{cp4} gives the time to compute the local maxima by one cluster.

$$t_{cp1} = \frac{3 \times t_{fl} \times N^2 \times \theta_c}{P} \tag{5.22}$$

$$t_{cp2} = \rho_c \times \theta_c \times \log_2 P_c \tag{5.23}$$

$$t_{cp3} = \frac{(n-1) \times \rho_c \times \theta_c}{P_c} \tag{5.24}$$

$$t_{cp4} = \frac{3 \times \rho_c \times \theta_c}{P_c}, \tag{5.25}$$

Intra-cluster and inter-cluster communication times are give by

$$t_{cl} = (\log_2 P_c + 1) \times \rho_c \times \theta_c \tag{5.26}$$

$$t_{icl} = \frac{n \times R \times P_p \times \theta_c \times \rho_c}{P_c}, \tag{5.27}$$

Similarly, the corresponding computation and communication times for PP are given by

$$t_{cp1} = \frac{3 \times t_{fl} \times N^2 \times (\frac{\theta_c}{n} + 2)}{P_c} \qquad (5.28)$$

$$t_{cp2} = \log_2 P_c \times \rho_c \times (\frac{\theta_c}{n} + 2) \qquad (5.29)$$

$$t_{cp3} = \frac{3 \times \rho_c \times \theta_c}{n \times P_c} \qquad (5.30)$$

$$t_{cl} = (\log_2 P_c + 1) \times (\frac{\theta_c}{n} + 2) \times \rho_c. \qquad (5.31)$$

Figure 5.21 depicts the speedups for hough transform using the two partitioning methods. Due to the communication overhead through global memory, which increases linearly with the number of clusters, the speedup for DP levels off. Figure 5.22 shows the computation and communication times for hough transform, whereas Figure 5.23 shows the communication overhead for hough transform in detail. Data partitioning does not perform as well as parameter partitioning. However, degradation with respect to best case speedup in DP is small. As we can observe, good speedup can be obtained for a global data dependent algorithm like hough transform. Figure 5.22 and 5.23 illustrate the computation and communication times for the DP case.

Figure 5.24 shows the speedup for hough transform (DP) and Figure 5.25 depicts the communication and computation times, respectively when the bus is used as a global interconnection network. Note that performance of the hough transform under PP will be the same in both cases because there is no global communication.

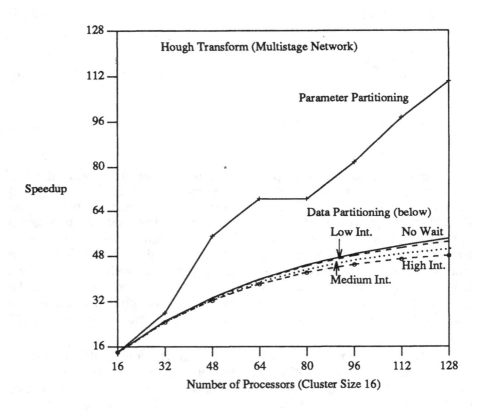

Figure 5.21 : Speedup for Hough Transform (Multistage Network)

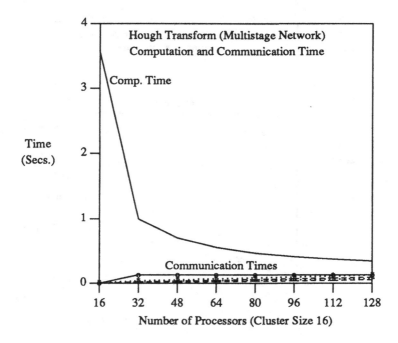

Figure 5.22 : Computation and Communication Times for Hough Transform

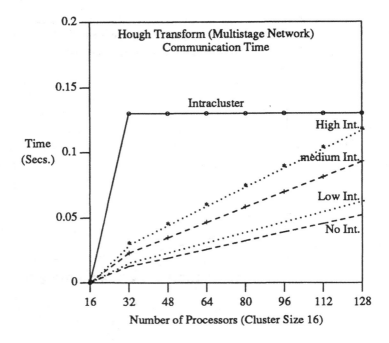

Figure 5.23 : Communication Times for Hough Transform

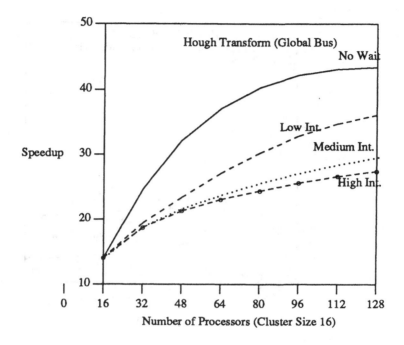

Figure 5.24 : Speedup for Hough Transform (Global Bus)

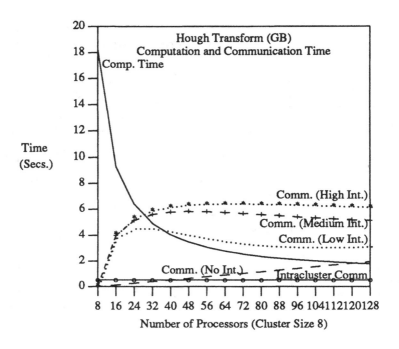

Figure 5.25 : Computation and Communication Times Hough Transform
        (Global Bus)

5.5. Summary

In this chapter we presented inter-cluster communication strategies in
NETRA. Inter-cluster communication is needed when algorithms are
mapped across clusters to transfer data between tasks executing on different
cluster processors, or to input and output data. Several factors contribute to
the performance of algorithms mapped across clusters. Not only does an
algorithm's computation and communication characteristics contribute to its
performance, but the system load, interference in accessing global interconnection network and global memory, and network bandwidth also contribute
to the performance. We presented an analysis of how the effect of conflicts in
global network and memory can be incorporated into the performance
evaluation of parallel algorithms mapped across clusters. For each algorithm
we presented one or more mapping strategies, its performance evaluation and
a discussion of the results.

The performance results were used to compare alternative inter-cluster
communication strategies, and they show that it is possible to obtain good

performance for algorithms with different characteristics under varying degrees of conflicts in a global interconnection network. In general, a multistage interconnection network as the global interconnection performs much better than a global bus, as expected. The parameters chosen for processor speed and communication speed were very conservative. We think that much faster processors and communication links are possible and available with today's technology, and therefore, the performance results presented in this chapter are also conservative. However, we obtained insight into the sensitivity of the performance measures as a function of various architecture parameters.

Chapter 6

Load Balancing and Scheduling Techniques

6.1. Need for Efficient Load Balancing Techniques

As discussed in Chapters 1 and 2, IVSs employ a sequence of image understanding algorithms in which the output of an algorithm is the input of the next algorithm in the sequence. Algorithms that constitute an integrated vision systems exhibit different characteristics, and therefore, require different data decomposition techniques and efficient load balancing techniques for parallel implementation. Since the input data of a task is produced as the output data of the previous task, this information can be exploited to perform knowledge based data decomposition and load balancing.

This chapter presents several techniques to perform static and dynamic load balancing schemes for IVSs. These techniques are novel in the sense that they capture the computational requirements of a task by examining the data when it is produced and using the knowledge of the computation in the next step. Furthermore, they can be applied to many integrated vision systems because many algorithms in different systems are either the same or have similar computational characteristics. These techniques are evaluated by applying them to the algorithms in a motion estimation system. It is shown that the performance gains when these techniques are used is significant, and the overhead of using these techniques is minimal. The evaluation is performed by implementing the algorithms on the hypercube multiprocessor which is a distributed memory system and Encore Multimax which is shared memory multiprocessor. The rationale behind using a commercially available machine is to capture all the overheads in implementations.

Data decomposition and load balancing techniques presented in this chapter are for medium to large grain parallelism. Two important characteristics of these techniques are that they are general enough to apply to any integrated system, and that they use statistics and knowledge from the execution of a task to perform load balancing and scheduling for the next task in the system. Furthermore, these techniques are architecture independent so that they can be applied to most MIMD machines. For example, in the motion estimation system sufficient knowledge can be obtained about the output data from the zero crossing step to perform efficient data decomposition and load balancing for the stereo match step. Knowledge from each step is used to perform load balancing in the next step. The advantages of such schemes are as follows. First, these techniques use characteristics of the tasks and the data, and therefore, work well no matter how the data changes. Secondly, many integrated vision systems consist of such tasks and exhibit the above described computation flow, and therefore, these techniques can be used in any system (e.g., object recognition, optical flow, etc.) .

Section 6.1 describes the proposed load balancing and data decomposition techniques. Section 6.2 presents a parallel implementation of these algorithms in an integrated environment and discusses the performance results for each of these algorithms, data decomposition and load balancing schemes. The underlying multiprocessor machines on which we have implemented these algorithms are intel iPSC/2 hypercube and Encore Mulimax. Some of these techniques have been applied to other integrated vision systems and have been shown to work well [57].

6.2. Load Balancing and Scheduling Techniques for Parallel Implementation

In a multiprocessor system an obvious and simple method to implement a task in parallel is to decompose the data and and the underlying tasks equally among the processors. In fact, parallelizing compilers parallelize DO loops in this fashion if some dependency criteria are satisfied. In a completely deterministic computation in which the computation is independent of the input data, such schemes perform well and normally the processing time is comparable on all the processors. That is, efficient utilization and load balancing can be obtained. For example, regular algorithms such as convolutions, filtering or FFT exhibit such properties. The amount of computation to obtain each output point is the same across all input data. Therefore, uniform decomposition of data results in load balanced implementation.

Most other algorithms do not exhibit the regular structure, and the computation is data dependent. Furthermore, the computation is not uniformly distributed across the input domain. In such cases, a simple decomposition does not provide efficient mapping and results in poor utilization and low speedups. Also, the performance cannot be predicted for a given number of processors and data size because the computation varies with the type of data and its distribution. For example, in the stereo match algorithm, the computation is more where the feature points are dense and is comparatively small where the number of features is small and sparsely distributed (Figure 2.4). Hence, uniformly partitioning the input data among processors is not expected to provide good speedups and utilization. One of the common approaches to parallelize tasks is divide-and-conquer. But the question arise what criteria should be used to divide a task? We propose techniques that use use the knowledge and characteristics of the IVSs to implement tasks using divide-and-conquer.

The following example (Figure 6.1) of a sequential computation, and its corresponding possible parallelization illustrates the concept. In Figure 6.1, if the computation in "Function" is data independent; that is, if it is uniform over the entire input domain, simple partitioning of indices for parallelization will work well. Therefore, values of $L_i(p)$, $L_j(p)$, $R_i(p)$, and $R_j(p)$ can be assigned such that each processor receives an equal amount of data. However, more important questions arise when "Function" is executed only when a certain condition is satisfied; for example, existence of an edge or value beyond a certain threshold. Then the above simple partitioning is not likely to distribute the computation equally among processors.

Consider the example shown in Figure 6.2 where, even though the granules can be distributed equally among processors, the computation exists only where "Condition" is satisfied. Therefore, the computation is likely to be distributed in a skewed manner. However, if it is possible to evaluate the condition long before execution of "Function", then using that knowledge, the computation can be appropriately partitioned among processors. That is, the values of the $L_i(p)$, $L_j(p)$, $R_i(p)$, and $R_j(p)$ can be computed using additional information, other than just granule size, that assigns partitions by equally distributing the computation. The above examples are very simple and are presented to illustrate the concept. In real vision applications the computations are much more involved. We need to investigate two important questions; 1) how do we incorporate that knowledge in parallel implementation, and 2) can we do it with little overhead? In the following we propose that we can incorporate such knowledge by exploiting characteristics of vision systems and that we can achieve it with little overhead. The basic approach is to evaluate the condition as soon as the corresponding data is computed (produced) rather than wait till it is used (consumed).

In an integrated vision system, it is important to efficiently allocate resources and perform load balancing at each step to obtain any significant performance gains overall. An important characteristic of such systems is that the input data of a task is the output of the previous task. Therefore, while computing the output in the previous task enough knowledge about

/* For an image of size $N{\times}N$*/

/*Code for processor p, $1{\leq}p{\leq}n$*/

```
FOR i=1 to N DO                   FOR i=L_i(p) to i=R_i(p) DO
FOR j=1 to N DO                   FOR j=L_j(p) to j=R_j(p) DO
 Im_out (i,j) = Func(Im_in (i,j))   Im_out (i,j) = Func(Im_in (i,j))
```

a) Sequential Code   b) Uniform Parallelization

Figure 6.1 : Parallelization for data independent computation

/* For an image of size $N{\times}N$*/

/*Code for processor p, $1{\leq}p{\leq}n$*/

```
FOR i=1 to N DO                   FOR i=L_i(p) to i=R_i(p) DO
FOR j=1 to N DO                   FOR j=L_j(p) to j=R_j(p) DO
 IF (Cond (Im_in (i,j) = true))     IF (Cond (Im_in (i,j) = true))
  Im_out (i,j) = Func(Im_in (i,j))    Im_out (i,j) = Func(Im_in (i,j))
```

a) Sequential Code   b) Parallelization

Figure 6.2 : Parallelization for data dependent computation

the data can be obtained to perform efficient scheduling and load balancing. In the following we discuss such techniques and in the next section we present the performance results for these techniques using algorithms in the motion estimation system.

Consider a parallel implementation of a task on an n processor parallel machine. Let T_i $(1{\leq}i{\leq}n)$ denote the computation time at processor node i. Then the overall computation time for the task is given by

$$T_{max} = \max\{T_1,...,T_n\} \qquad (6.1)$$

The total wasted time (or idle time) T_w is given by

$$T_w = \sum_{i=1}^{i=n}(T_{max} - T_i) \qquad (6.2)$$

If $T_{max} = T_i$ for all i, $1 \le i \le n$, then the task will be completely load balanced. Another measure of imbalance is given by the variation ratio V,

$$V = \frac{T_{max}}{T_{min}}, \quad T_{min} = \min\{T_1,...,T_n\} \qquad (6.3)$$

The goal in performing load balancing is to minimize T_w, or move V as close to 1 as possible. In the best case, $T_w = 0$ or $V_1 = 1$. If T_{seq} is the time to execute the same task on a sequential machine, then the speedup is given by

$$S_p = \frac{T_{seq}}{T_{max}} \qquad (6.4)$$

Therefore, by minimizing T_w, the achievable speedup can be maximized.

6.2.1. Uniform partitioning

Data decomposition using uniform partitioning performs well as a load balancing strategy for input data independent tasks because equally dividing the data distributes the computation equally. If the total input data size is D then the total computation time to execute the task is $T = k \times D$, where k is a determined by the computation at each input data point. For example, in convolution of a image with an $m \times m$ kernel, $k = 2 \times m^2$ floating point operations. Hence, for an n node multiprocessor the data decomposition methods to balance the computation is to make the granule size to

$$d_i = \frac{D}{n} \qquad (6.5)$$

For data independent algorithms, such a partitioning guarantees equal distribution of computation among processors. Therefore, if communication time can be minimized, then optimal performance can be obtained on a given multiprocessor.

6.2.2. Static scheduling (First-order scheduling)

When the computation is not uniformly distributed across the input domain and is data dependent then uniform partitioning does not work well for load balancing. Normally, the computation depends on the significant data or the type of data in a partition. Many image processing and vision algorithms exhibit this behavior. For example, in stereo match, and hough transform the computation is proportional to the number of features (edges) or significant pixels in a granule rather than on the granule size. Therefore, equal size granules do not guarantee load balanced partitioning because of the data dependent nature of the computation. In fact, the variation can be very significant as we shall observe in the next section when we discuss the performance. In many such algorithms, the computation time for a granule (i), T_i, is proportional to a certain extent on the granule size (fixed overhead to process a granule) and to the number of significant data in the granule. That is,

$$T_i = A \times d_i + B \times f_i \qquad\qquad (6.6)$$

where, d_i is the granule size, f_i is a measure of significant data in a granule (i), and A and B are arbitrary constants which depend on the algorithm. Therefore, the objective is to divide the computation among the processors such that each processor receives an equal measure of computation to perform rather than an equal amount of data. One way to assign a granule to a processor will be to compute the total measure of computation and partition as follows:

$$T_i = \frac{\sum\limits_{i=i}^{i=g} A \times d_i + B \times f_i}{n} \qquad\qquad (6.7)$$

where g is the total number of granules in the input domain (Note that the number of granules for the current task is n for an n processor system).

For example, consider computing hough transform of an edge image. The algorithm involves computing the parameters for line segments in the images. If there exists a line whose normal distance from the origin is r, the normal makes an angle θ with the x-axis; then if the point (x,y) lies on that line, the following Equation is satisfied.

$$r = x cos\theta + y sin\theta$$

r and θ are quantized for desired accuracy and then for each significant pixel (where there is an edge), and r is computed for all quantized θ values. If two partitions of equal size contain a different number of edge pixels, then the amount of computation will be different for the two partitions despite their being equal in size. In fact, the computation is directly proportional to the number of edge pixels in the partition. A way to perform static load balancing will be to decompose the input data such that each partition contains an equal number of edge pixels. The computation to recognize this partioning can be performed in the task in which edges are detected by keeping a count of the number of edges detected by a processor. Note that it is important to compute the statistics on the fly when edges are detected to guarantee low overhead. If the same statics are gathered by sequentially scanning the input data then the overhead can be significant. Once the task is completed, the data can be reorganized such that the number of edges with each processor is in the interval $(\frac{Z_a}{n} - \delta, \frac{Z_a}{n} + \delta)$, where Z_a is the total number of edges detected in the image and δ is determined by the minimum granule size from fixed overhead considerations.

6.2.3. Weighted static scheduling (Second-order scheduling)

When the computation in a granule not only depends on the number of significant data points in the input domain but also depends on their spatial relationships, then data distribution also needs to be taken into account as a measure of load to perform load balancing. For example, from the previous section it is evident that to perform stereo match, not only does the computation depend on the number of zero crossings but also depends on their spatial distribution. If the zero crossings are densely spaced then the computation will be more than that if the same number of zero crossings is sparsely distributed (refer to Figure 6.2). The reason is that if the zero crossings are densely packed then greater numbers of zero crossings need to be matched with each corresponding zero crossing in the other image, whereas fewer numbers of zero crossings need to be matched if they are sparsely distributed. Hence, the computation also depends on the spatial density (such as features/row if one-dimensional matching is performed). That is,

$$T_i = A \times d_i + B \times w_i \times d_i \tag{6.8}$$

where w_i is the feature dependent spatial density. For example, if the minimum granule size is a row of the input data then $w_i = r_i^\beta$, where r_i is the number of features in row i and β is a parameter, $0 \leq \beta \leq 1$. $\beta = 0$ means that the computation is independent of how the features are distributed within a row. Therefore, to divide the computation equally among n processors, the following heuristic can be used.

$$T_i = \frac{\sum_{i=0}^{i=R} A \times d_i + B \times w_i \times d_i}{n} \tag{6.9}$$

where, R is the number of rows in the image. Note that the above heuristics approximate the load and do not exactly divide the computation among processors. However, in the next section we will show that these schemes perform well.

As an example, consider the stereo match computation. While partitioning the data among processors, a weight can be assigned to each row as a function of the number of features in the row. This weight represents the feature density. Note that using a row as the smallest granule avoids the communication overhead because search space for stereo matching is one-dimensional, and therefore, if the granule boundary is one row then there is no need for communication.

6.2.4. Dynamic

The above three methods use the knowledge about the data when it is produced to perform load balancing for the next task. However, once decomposition is done then the data is not reshuffled. Therefore, we consider the above methods as knowledged based static load balancing schemes. In the dynamic scheme, the data is decomposed into finer granules such that the number of tasks (that is, the number of independent granules) M is much larger than the number of processors.

At execution time, the processors are assigned these tasks dynamically by a designated scheduler from a task queue which contains these tasks. Processors are assigned new tasks as they finish their assigned tasks, if there are more tasks left to be assigned. However, the knowledge obtained from the previous step again can be used to anticipate the completion of a task to assign a new task to a processor. That is, the tasks can be pipelined, and therefore, the overhead of the dynamic load balancing can be reduced. The communication overhead of dynamically assigning tasks is not incurred in the previous three schemes.

The following procedure in Figure 6.3 illustrates the dynamic assignment of tasks onto the processor. The pseudo code essentially illustrates what the scheduler does in order to perform dynamic load balancing. The number of tasks (max_tasks) are determined during the execution of the preceding step in the system, and the task_queue contains all the tasks including the computational information associated with each task. Initially, the scheduler assigns few tasks to each processor. The number of tasks to be assigned initially is a parameter (pipe_line_no). If this parameter is 1, it implies that there is no anticipatory scheduling. In other words, a processor is assinged a new task only when it finishes the task it is currently executing. A task is assigned to a processor only if the task contains significant computation. For example, in stereo match, if a task's data does not contain any zero crossings,

Dynamic Scheduling of Tasks

/*Initial Assignment*/

```
1.      curr_task = 0;
2.      for j = 1 to j <= pipe_line_no do
3.              for i = 1 to i = num_proc do
4.                      if comp(task_queue(curr_task)) > 0
5.                              schedule curr_task at proc. P_i;
6.                              curr_task = curr_task+1;
7.                      else
8.                              curr_task = curr_task+1;
9.                              go to 4.
10.                     end_if
11.             end_for
12.     end_for
```

/*Scheduling*/

```
13.     done = false; k = num_proc;
14.     while not done do
15.             wait for msg from a processor;
16.             receive msg;
17.             if ( msg = compl_msg )
18.                     P_i = sender processor;
19.                     if curr_task < max_tasks
20.                             if comp(task_queue(curr_task)) > 0
21.                                     schedule curr_task at proc. P_i;
22.                                     curr_task = curr_task+1;
23.
24.                             else
25.                                     curr_task = curr_task+1;
26.                                     go to 19.
27.                     else
28.                             send term_msg to P_i.
29.             else if ( msg = term_msg)
30.                     k = k - 1;
31.                     if (k <= 0)
32.                             done = true.
```

Figure 6.3 : Dynamic Scheduling

then the task can be discarded because it is not going to produce any useful information anyway. In a blind scheme, where little is known about a task, the task will be assigned, which is an overhead, and can be avoided by using the knowledge obtained from the previous steps. Whenever a processor P_i completes the current task, it sends a *compl_msg* to the scheduler which assigns P_i a new task if the task_queue is not empty. Once the task_queue becomes empty, the scheduler sends a *term_msg* (terminate message) to all the processors. Upon receiving a *term_msg* from the scheduler, processors complete the remaining tasks in their task_queues, and sends a *term_msg* to the scheduler, terminating the computation. Note that by using the pipe_line_no, anticipatory dynamic scheduling can be performed, and a processor need not be idle when a new task is being assigned. By using this parameter, the amount of initial static assignment, and dynamic assignment can be controlled. In a shared memory machine, each task participate in dynamic scheduling using a shared array of available tasks. Upon completing the current task, a processor locks the task array and takes the next available significant task. Figure 6.4 shows the partitioning for the above described strategies for the stereo match algorithm.

6.3. Parallel Implementation and Performance Evaluation

This section presents a parallel implementation of the algorithms that are part of the motion estimation system and describes the performance of the algorithms and load balancing strategies.

6.3.1. Feature extraction

Features used for stereo match algorithms are the zero crossings of the convolution of the image with Laplacian, as presented in Section 6.1. Zero crossing computation involves 2-D convolution and extraction of zero crossings from the convolved image. Since convolution is a data independent algorithm, uniform partitioning is sufficient to evenly distribute the computation. The mapping is a division of $N \times N$ image onto P processors. Each processor computes the zero crossings of a share of N^2/P pixels (Equation 6.5). Data division onto the processors is done along the rows. This mapping reduces communication to only one direction. The reason is as follows. 2-D convolution can be broken into two 1-D convolutions [48]. This not only reduces the computation from W^2 sum of products operations per pixel to $2 \times W$ sum of product operations per pixel (W is the convolution mask window size) but also reduces the communication requirements in a parallel implementation if the data partitioning is done along the rows. There is no need for

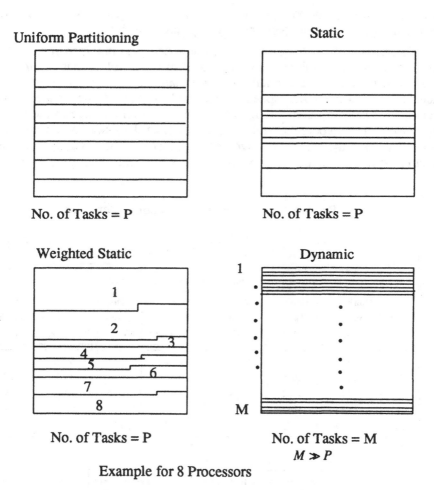

Example for 8 Processors

Figure 6.4 : Load Balancing Strategies

communication when convolution is performed along the rows.

Table 6.1 shows the performance results for the above implementation for an image of size 256×256 and a convolution window of size 20×20. The first column shows the number of processors in the cube(P). The second column represents the total processing time (t_{proc}) for convolution. Column 3 shows the number of bytes communicated by a processor to the neighboring processor, and column 4 shows the corresponding communication time which is small compared to the computation time. The second half of the table shows computation time for extracting zero crossings from the convolved image. The corresponding speedups are also shown.

It can be observed that almost linear speedup is obtained for convolution. The two factors which contribute towards this result is that communication overhead is relatively small and is constant as the number of processors increases. However, the speedup obtained in the elapsed time (which includes the program and data load time also) is sublinear due to the following reason. The hypercube multiprocessor's host does not have a broadcast capability, and therefore, the overhead of loading the program increases linearly with the number of processors. However, data load time increment

Table 6.1 : Performance for Feature Extraction (Zero Crossings)

Computation for Convolution and Zero Crossings							
	Convolution Window Size = 20x20						
No. Proc.	Conv. Comp. Time(sec.)	Conv. Comm. Bytes	Conv. Comm. Time(ms.)	Conv. Total Time(sec.)	Conv. Speed Up	ZC Comp. Time(sec.)	ZC Speed Up
1	109.0	0	0	109.0	1	6.47	1
2	54.76	2816	13	54.78	1.98	3.23	1.99
4	27.51	5632	36	27.55	3.95	1.66	3.89
8	13.88	5632	36	13.92	7.83	0.85	7.60
16	7.07	5632	36	7.11	15.33	0.42	15.25

Feature Extraction Performance (Elapsed Time)		
No. Proc.	Elapsed Time(sec.)	Speed up
1	116.2	1
2	58.8	1.97
4	30.1	3.86
8	16.1	7.22
16	9.6	12.1

with the increase in the number of processors is comparatively small because amount of data to be loaded to one processor decreases as the number of processors increase. The only increment in data load time results from the number of communication setups from the host to the node processors which increases linearly with the number of processors.

Figure 6.5 shows the speedup obtained for zero crossing computation when implemented on a shared memory (Encore Multimax) machine. The figure illustrates that almost linear speedup can be obtained. Note that since zero crossing involves convolution and thresholding which are data independent algorithm, a uniform decomposition is sufficient to distribute the load evenly among processors.

6.3.2. Matching features

This task involves matching features in stereo pairs of images. As discussed in Chapter 2, the epipolar constraint limits the search for a match in

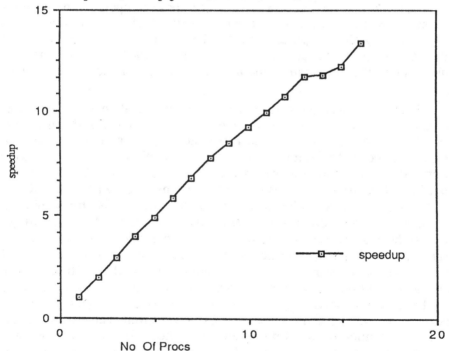

Figure 6.5 : Speedup for Zero Crossing Computation on Encore Multimax

the corresponding image to only in horizontal direction, i.e., along the rows in the zero crossings of the image. Thus data partitioning along the rows for parallel implementation results in no communication between node processors as long as each partition contains an integral number of rows.

The computation involved in the stereo matching algorithm is data dependent and varies across the image because it depends on the number of zero crossings, distribution of zero crossing across the image and distribution of zero crossings along the epipolar lines. Therefore, partioning the data uniformly among the processors (i.e., assign each processor an equal number of rows) may not yield expected speedups and processor utilization. A processor which has very few zero crossings and sparsely distributed zero crossings will be under utilized whereas a processor with a large number of zero crossings and densely distributed zero crossings will become a bottleneck, and this imbalance of load will result in a poor performance.

We used uniform partitioning, static load balancing, weighted static and dynamic load balancing schemes to decompose computation on the multiprocessor. Static load balancing can be achieved by keeping a count of the zero crossings with each processor when the previous task (convolution and feature extraction) is executed. At the completion of the task, the data is reorganized, using this information and the techniques described in the previous section.

Figure 6.6 shows the distribution of the computation times for an 8-processor case. The X-axis shows the processor number and the Y-axis shows the computation time for each scheme. As we can observe, uniform partitioning does not perform well because the variation in computation time is tremendous, and therefore, performance gains are minimal. The static load balancing scheme (shown as dashed bars) performs much better than uniform partitioning, but the variation in the computation times is still significant because the computation also depends on the distribution of zero crossings. The weighted static scheme does perform better than static and further reduces the variation in computation times. Note that these schemes only measure the load approximately, and therefore, will not divide the computation exactly uniformly. Furthermore, minimum granularity is a row boundary in order to avoid communication between processors. Finally, for an 8-processor case, the dynamic scheme performs very well. Table 6.2 summarizes the distribution for the 8-processor case. The table shows the computation time for each processor for all four methods. Speedup is computed as follows. If T_s is the sequential processing time and T_{max} is the maximum processing time of one processor among n processors, then speedup is $\frac{T_s}{T_{max}}$.

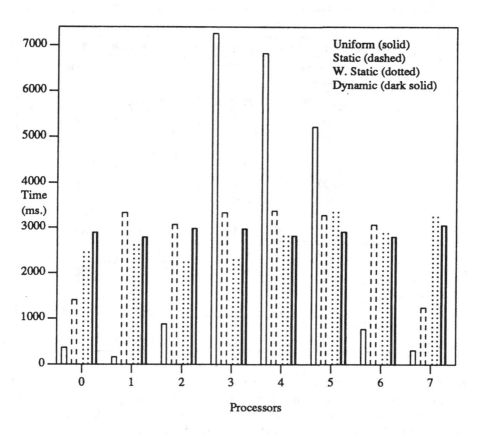

Figure 6.6 : Distribution of Computation Times (Stereo Match)

Table 6.2 : Distribution of Computation Times for Stereo Match

Computation Time Distribution for Stereo Match (P=8)				
Proc. No.	Uniform Partitioning	Static	Static Weighted	Dynamic
	Time (ms.)	Time (ms.)	Time (ms.)	Time (ms.)
0	364	1402	2439	2890
1	164	3333	2606	2786
2	878	3066	2219	2980
3	7258	3327	2277	2967
4	6827	3371	2798	2818
5	5207	3269	3328	2913
6	762	3063	2864	2803
7	312	1243	3223	3051
Max.	7258	3371	3328	3051
Min.	164	1243	2219	2786
Variation ratio	44.25	2.71	1.50	1.09
Improvement ratio	1	2.15	2.19	2.38

Variation ratio is the ratio of the maximum processing time to the minimum processing time and it provides a measure of imbalance in the computation. For example, in Table 6.2 the variation ratio is 44.25 for the case of uniform partitioning, 2.71 for the case of static load balancing, 1.50 for weighted static and 1.09 for dynamic load balancing. Improvement ratio is the ratio of speedup obtained with load balancing to that of uniform partitioning. The computation times shown in these tables include all the overhead of load balancing schemes. Figure 6.7 shows the speedup graph for varying sizes of multiprocessors from 1 processor to 16. As we can observe, uniform partitioning does not provide any significant gains in speedup as the number of processors increases. Dynamic scheme performs the best among all the schemes (at least for small processor size) but the two static schemes perform comparable to the dynamic scheme. We believe that as the number of processors is increased, the two static schemes will move even closer to the dynamic scheme or even perform better than the dynamic scheme because for larger multiprocessors, the overhead of the dynamic scheme will be larger. One important conclusion from the above observations is that such a knowledge based scheme performs very well to schedule parallel tasks in an

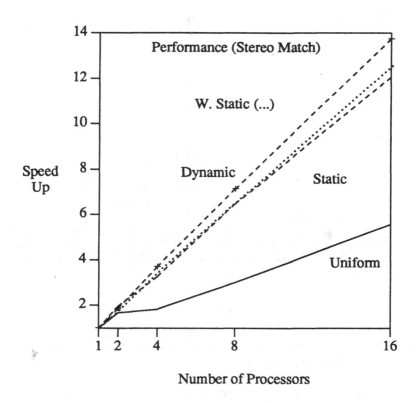

Figure 6.7 : Speedups for Stereo Match Computation

integrated vision system in which very often similar bottom up computations are performed in a sequence.

Figure 6.8 shows the distribution of stereo match computation time for an 8-processor Encore Multimax. Clearly, in this case also, uniform partitioning does not perform well at all, and computation times are highly skewed. Static scheme (first-order) performs much better than uniform partitioning and computation times are more evenly distributed. Weighted-static (second-order) performs marginally better than static scheme. Finally, dynamic scheme performs the best among all the schemes.

The summary of computation times for multiprocessors from sizes 1 to 16 is depicted in Figure 6.9 for stereo match computation on Encore Multimax. The skew effect is clearly visible for computation times for uniform

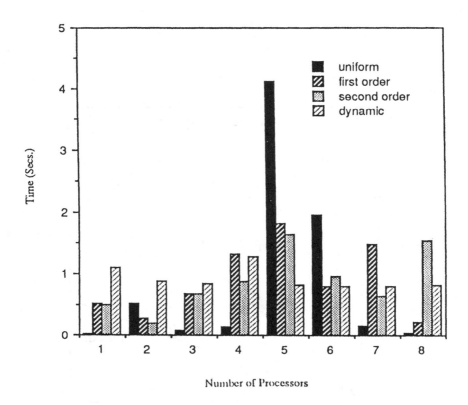

Figure 6.8 : Distribution of Computation Times for Stereo Match (P=8)
(Encore Multimax)

partitioning scheme. Since the computation is highly data dependent, increasing the number of processors may increase the computation time as illustrated in Figure 6.9. Upto 16 processors, the dynamic scheme consistently performs the best. Static and weighted static schemes perform comparably. However, it is not clear why weighted scheme performs worse than static when the number of processors is increased beyond 13.

Figure 6.9 also shows the overhead of different scheme, denoted as "TH creation time" in the figure. The time includes thread (which is a process in Encore OS) creation time, i.e., how long it takes to create the number of desired processes. In summary, both for hypercube and Encore Multimax implementations, we observe that the proposed schemes perform well and much better than uniform partitioning. The results highlight and reinforce the claims that these techniques are architecture independent and can be used in different MIMD machines with very little overhead with significance performance improvements.

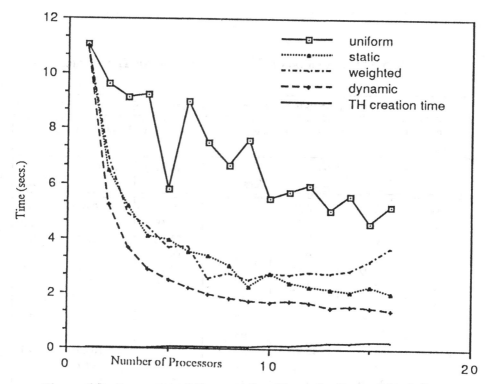

Figure 6.9 : Summary of Computation Times for Stereo Match for
various Multiprocessor Sizes (Encore Multimax)

6.3.3. Time match

The computation in the time match algorithm is similar to that in stereo
match except the search space is two-dimensional and the input to the algorithm is the stereo match output. Another difference is that the number of
significant points in the input data is much smaller than that in stereo match
because a great deal of input points get eliminated in stereo match. Table 6.3
shows the distribution of the computation times for the 16 processor case.
We only present uniform partitioning and static load balancing cases. The
most important observation is that uniform partitioning performs worse than
that in the case of stereo match and static load balancing performs better.

The table shows how the measure of computation (number of zero
crossings left from stereo match step) gets divided among the processors in
the two cases. It is clear that the number of zero crossings is very evenly distributed (within the minimum granule of one row constraint) in the static
case whereas they are lumped with a few processors in the uniform

Table 6.3 : Distribution of Computation Time for Time Match

Computation for Time Match (Proc. = 16)					
Proc. No.	Uniform Partitioning			With Load Balancing		
	Matching (Sec.)	Total (Sec.)	No. Zcs	Matching (Sec.)	Total (Sec.)	No. Zcs
0	0.14	0.22	3	9.35	10.00	47
1	0.03	0.14	2	12.38	12.55	50
2	0.02	0.13	0	13.12	13.21	53
3	0.02	0.13	0	14.23	14.27	43
4	0.02	0.13	0	11.88	11.91	45
5	3.61	3.72	21	10.93	10.95	44
6	13.45	13.56	55	12.82	12.85	53
7	5.09	5.20	20	12.16	12.19	51
8	26.65	26.76	93	11.41	11.44	45
9	45.85	45.97	182	10.63	10.65	40
10	73.82	73.93	259	13.89	13.91	50
11	27.20	27.32	121	13.69	13.71	44
12	0.31	0.42	3	15.07	15.09	43
13	0.11	0.22	1	15.70	15.72	56
14	0.42	0.53	4	14.36	14.39	56
15	0.08	0.10	0	5.21	5.68	43
	Max. time(sec.)	Min. time(sec.)	Variation ratio	Speed up	Improvement ratio	
Uniform	73.82	0.10	738	2.69		
Balanced	15.72	5.68	2.76	12.63	4.7	

partitioning case. Figure 6.10 shows the speedup graphs for the two schemes for a range of multiprocessor size. The speedup gains for the load balanced case is very significant over the uniform partitioning case. We computed the overhead of performing knowledge based static load balancing and the overhead was 3 ms., which is negligible compared to the computation time, and the performance gains are significant.

6.3.4. Second stereo match

This step involves stereo match computation for features from images at time instant t_{i+1} after time point correspondence is established between images at time t_i and t_{i+1}. The matching is similar to that in the first stereo match except that it needs to be done only at those points at which time correspondence has already been established. Consequently, the number of features to be matched is much less than that in the first computation, and hence, the importance of load balancing is further increased. Figure 6.11 depicts the distribution of computation times for the second stereo match step. The three load balancing algorithms used in this case are Uniform

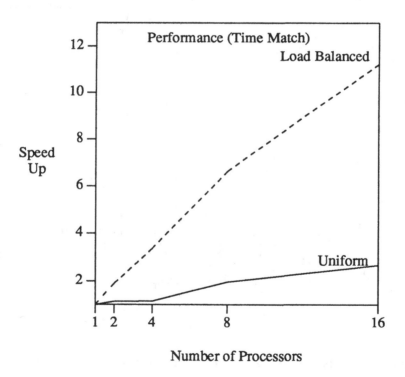

Figure 6.10 : Speedup for Time Match

Partitioning, Static and Dynamic. As it is observed from the Figure, the uniform partitioning does not perform well at all compared to the other two schemes. The variation in computation time is significant. Furthermore, it is observed that static and dynamic schemes perform comparably.

Figure 6.12 presents the speedups for the same algorithm for various multiprocessor sizes. The Figure shows that the gains from these load balancing schemes are very significant over simple uniform partitioning. One

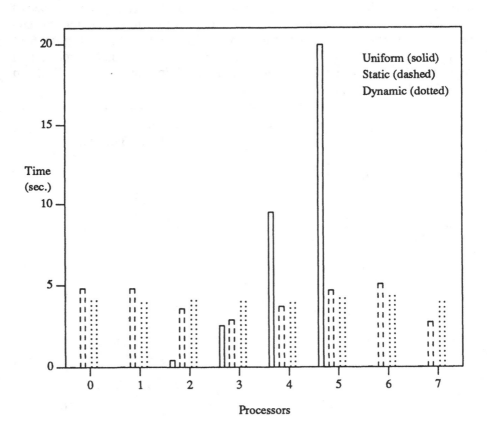

Figure 6.11 : Distribution of Computation Times for Second Stereo Match (P=8)

important observation can be made by comparing results in Figure 6.7 and 6.12. Note that the performance of uniform partitioning in the second stereo match is much worse than that in the first stereo match. For example, for the 16 processor case, the speedup in the first case is 5.55, whereas for the same multiprocessor size speedup is only approximately 2.3. Therefore, as the computation progresses in an integrated environment, the gains of these load balancing schemes become increasingly significant. Hence, overall gains for the entire system are better than what may be expected.

6.3.5. Summary

In summary, the following important observations can be made from all the results presented in the previous sections. First, the improvement in

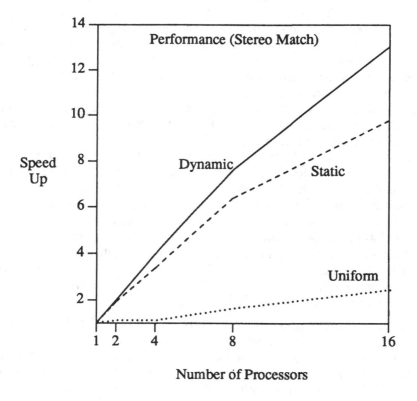

Figure 6.12 : Speedups for Second Stereo Match

performance (such as utilization and speedup) itself increases using the load balancing schemes as the number of processors increases. Therefore, performance gains are expected to be higher for larger multiprocessors. Secondly, in an integrated environment, the overheads of such methods are small because the measure of loads can be computed on the fly as a side result of the current task. Finally, though we showed the performance results of the implementation on the hypercube multiprocessor, these methods can be applied when algorithms are mapped on any medium to large grain multiprocessor system because these techniques are independent of the underlying multiprocessor architecture.

Consider the overall performance gains for the entire system. As the computation progresses from one step to the next, uniform partitioning performs worse because the data points reduce, but the computation at each point increases. Therefore, the gains of using parallel processing are minimal. However, the load balancing techniques recognize the data distribution at each step and data is decomposed using the distribution. Therefore, performance gains are expected to improve as the computation progresses in an integrated systems environment. For example, consider zero crossing, stereo match and time match and second stereo match steps. In zero crossing computation, uniform partitioning performs well and the load is balanced. Hence, the improvement ratio is 1. For stereo match the improvement of static over uniform partitioning is 2.15 for 8 processor case, and is 2.22 for the 16 processor case. Similarly, for the time match step, the improvement of static load balancing for 8 processor case is 3.38 and for the 16 processor case it is 4.2. Furthermore, for the second stereo match step, similar results are obtained. Therefore, it is observed that the improvement in performance itself increases as the number of processors increases as well as when the computation progresses in an integrated vision system. In summary, the performance gains are expected from these schemes for data decomposition and load balancing schemes as the number of processors increases, as the computation progresses in an integrated environment, and the overhead of these schemes is negligible compared to the performance gains.

Finally, through implementation on two completely different architectures such as the hypercube which is a distributed memory multiprocessor and Encore Multimax which is shared memory, bus-based multiprocessor it shown that the proposed schemes for load balancing and scheduling perform really well. There are several refinements possible to these scheme including the one in which static and dynamic schemes can be involved to produce better performance improvements.

Chapter 7

Concluding Remarks

7.1. Summary and Discussion

This book has addressed several issues in multiprocessor architectures and parallel algorithms for Integrated Vision Systems. The approach has been to consider computational requirements for vision applications in an integrated environment in designing a multiprocessor architecture rather than to propose architecture solutions to perform individual algorithms efficiently. An IVS involves algorithms from several levels of processing and the characteristics of algorithms in each level differ tremendously from algorithms in other levels. However, these algorithms need to exist in a system simultaneously and interact with each other. Therefore, a multiprocessor architecture suitable for IVS applications must be partitionable and reconfigurable. It must have the capability to allocate resources dynamically, provide for dynamic load balancing and task scheduling, provide fast and flexible communication, provide efficient I/O, and be fault-tolerant, in addition to providing raw processing power.

We presented a model of computation for IVSs. The model captures the computation requirements in an IVS, spatial as well as temporal data dependencies, and suggests what types of parallelism may be available in tasks of an IVS. Using the model desired features and capabilities of an architecture suitable for IVSs are identified. Another important aspect of the model is that it incorporates temporal relationships between tasks which are absent in the Image Understanding Benchmarks presented in [1].

A multiprocessor Architecture for IVSs (called NETRA) was presented. The original form of NETRA was proposed by Sharma, Patel and Ahuja in [3]. Several refinements to the architecture have been presented in this book after careful and detailed considerations of the computational requirements of an IVS. The modifications include alternative inter-cluster communication strategies and the synchronization bus in a cluster. The architecture was critiqued in the light of computation requirements for IVSs developed earlier. It was argued that the architecture provides most of the features, such as reconfigurability, partitionability, flexible communication, and fast I/O, needed in a multiprocessor of IVS. Furthermore, a discussion on how to provide these capabilities was included.

Performance of various algorithms on a processor cluster was presented. The evaluation of a cluster using several algorithms indicates that the cluster provides flexibility of communication, ability to reconfigure in SIMD, MIMD and systolic modes, and shown that almost linear speedups are possible in most cases. The most important observation is that the programmable crossbar design reduces the overheads of mapping parallelism by providing selective broadcast capability and the ability to provide the best interconnection for a particular algorithm. Both analytical and implementation results were presented. Performance evaluation of some algorithms from the Image Understanding Benchmark were also presented.

We presented alternative inter-cluster communication strategies in NETRA and evaluation of parallel algorithms when mapped across multiple clusters. When an algorithm is mapped on multiple clusters, processors between different clusters need to communicate. This requires accessing the global interconnection network and global memory. However, conflicts may occur in accessing global interconnection and global memory, which in turn, affects the performance of an algorithm. Presented in the first part are two inter-cluster communication strategies, viz; using global memory via multistage network and a high speed bus connecting the clusters together. An analysis was made to compute delays through the global network due to conflicts so that their effects can be incorporated into performance evaluation of algorithms. In the second part, performance of several algorithms, when mapped across multiple clusters, is presented. The results indicate that even in the case of a large number of conflicts, good (almost linear speedups) performance can be obtained for several algorithms when a multistage network is used. However, in order for the bus to be a viable global interconnection, the bus bandwidth and speed must be much greater than the processor speed.

Data decomposition and load balancing techniques for implementing IVSs were presented. In order to obtain any significant performance gains from parallel implementation of intermediate and low level algorithms, efficient load balancing is important because the computation is normally data dependent. The main contributions have been to present techniques to perform data decomposition and load balancing schemes that exploit knowledge about the computation and the data in a task. Since in an IVS such knowledge for the next task is normally available while performing the current task, the overheads are minimal. Four techniques presented are uniform partitioning, which is shown to be good enough for data independent algorithms: static and weighted static, which are shown to perform well when computation is dependent on the amount of significant data and its distribution; and the dynamic load balancing which is shown to work as well. However, in the dynamic load balancing scheme, the spatial relationships between data element are not maintained. Performances of all the techniques have been shown using algorithms from a motion estimation system, and it is concluded that these schemes perform well and provide tremendous improvements in utilization and speedups. Finally, through implementation on different multiprocessor architectures the proposed load balancing and scheduling techniques were shown to be architecture independent.

7.2. Extensions

Future work relating to what is presented in this book can be put into three categories, namely, architectural issues, parallel algorithms issues and systems issues. The following is a brief discussion on each of these issues.

Specification and more detailed design of the communication networks in NETRA are areas for an extension to the current work. This involves design of communication protocols for both intra-cluster and inter-cluster communications. For example, the design of communication protocols should address issues such as where does the responsibility of programming the crossbar lie, or how the crossbar will be addressed. Once the specifications are provided, the crossbar can be designed in detail. Another task is to develop a versatile simulator for NETRA which incorporates clusters as well as other parts of NETRA in detail. The simulator can be used to evaluate algorithms in detail, and using the results of the evaluation, refinements in the architecture can be performed.

The design of NETRA also need not follow exactly the structure presented in this book. Some of the clusters can be special purpose multiprocessors to perform specific and high-speed computations (for example, FFT, convolution). Different clusters can be of different sizes depending on the requirments. Furthermore, some clusters can be replaced by available multiprocessors that perform well for a range of algorithms

The second avenue for future research is in the area of parallel algorithm issues, both specific to implementing them on NETRA as well as on other architectures. The foremost task is to map a variety of algorithms to evaluate the architecture as well as to develop a general approach to mapping parallel algorithms. Study and evaluation of algorithms, especially intermediate and high level, are necessary to obtain a better understanding of their characteristics from parallel implementation perspective. Furthermore, if the algorithms are evaluated using existing parallel machines, then a better understanding obtained for architectural issues as well as knowledge can be gained about overheads associated with implementation of such algorithms. The global shared memory structure can be utilized to store eand manage a knowledge base that contains information regarding mapping and performance of different algorithms on various topologies.

The third issue, termed systems issue, deals with the design of operating systems and development of a programming environment for NETRA. The design should specify tasks for various elements of the architecture, how the tasks interact with one another, how to specify and incorporate data and knowledge base about vision systems, design of protocols to use data, and knowledge base. Furthermore, the systems issues include designing protocols for load balancing, partitioning and allocating resources, memory management and task management. The most important and ultimate task is to integrate all of the above into one system.

References

[1]   C. Weems, A. Hanson, E. Riseman, and A. Rosenfeld, "An integrated image understanding benchmark: recognition of a 2 1/2 D mobile," in *International Conference on Computer Vision and Pattern Recognition*, Ann Arbor, MI, June 1988.

[2]   A. Choudhary and J. H. Patel, "A parallel processing architecture for integrated vision systems," in *17th Annual International Conference on Parallel Processing*, St. Charles, IL, pp. 383-388, August 1988.

[3]   M. Sharma, J. H. Patel, and N. Ahuja, "NETRA: An architecture for a large scale multiprocessor vision system," in *Workshop on Computer Architecture for Pattern Analysis ans Image Database Management*, Miami Beach, FL, pp. 92-98, November 1985.

[4]   J. L. Bentley, "Multidimensional divide-and-conquer," *Communications of the ACM*, vol. 23,, pp. 214-229, April, 1980.

[5]   M. J. B. Duff, "CLIP 4: a large scale integrated circuit array parallel processor," *IEEE Intl. Joint Conf. on Pattern Recognition*, pp. 728-733, November 1976.

[6]   M. J. B. Duff, "Review of the CLIP image processing system," in *National Computer Conference*, Anaheim, CA, 1978.

[7]   L. Cordella, M. J. B. Duff, S. Levialdi, "An analysis of computational cost in image processing: a case study," *IEEE Transactions on Computers*, vol. c-27, no.10, pp. 904-910, 1978.

[8]   Arvind, D. K. Robinson, and I. N. Parker, "A VLSI chip for real-time image processing," *IEEE International Symposium on Circuits and Systems*, pp. 405-408, 1983.

[9]   R. Davis and D. Thomas, "Geometric arithmetic parallel processor-systolic array chip meets the demands of heavy duty processing," *Electronic Design*, pp. 207-218, October 1984.

[10]   K. Batcher, "Design of a massively parallel processor," *IEEE Transactions on Computers*, vol. 29, pp. 836-840, 1980.

[11]   T. Kushner, A. Y. Wu, and A. Rosenfeld, "Image processing on MPP:1," *Pattern recognition*, vol. 15,, pp. 120-130, 1982.

[12]   J. L. Potter, "Image processing on the massively parallel processor," *IEEE Computer*, pp. 62-67, January 1983.

[13]   V. Cantoni, S. Levialdi, M. Ferretti, and F. Maloberti, "A pyramid project using integrated technology," in *Integrated Technology for Parallel Image Processing*, London, pp. 121-132, 1985.

[14]   A. Merigot, B. Zavidovique, and F. Devos, "SPHINX, A pyramidal approach to parallel image processing," *IEEE Workshop on Computer Architecture for Pattern Analysis and Image Database Management*, pp. 107-111, November 1985.

[15]   D. H. Schaefner, D. H. Wilcox, and G. C. Harris, "A pyramid of MPP processing elements - xperience and plans," *Hawaii Intl. Conf. on System Sciences*, pp. 178-184, 1985.

[16]   S. L. Tanimoto, "A hierarchical cellular logic for pyramid computers," *J. of Parallel and Distributed Processing*, vol. 1, pp. 105-132, 1984.

[17]   S. L. Tanimoto, T. J. Ligocki, and R. ling, "A prototype pyramid machine for hierarchical cellular logic," in *Parallel Hierarchical Computer Vision, L. Uhr (Ed.)*, London, 1987.

[18]   N. Ahuja and S. Swamy, "Multiprocessor pyramid architectures for bottom-up image analysis," *IEEE Transactions on Pattern Analysis and Machine Intelligence*, vol. PAMI-6, pp. 463-475, July 1984.

[19]   A. Rosenfeld, "The prisom machine: an alternative to the pyramid," *Journal of Parallel and Distributed Computing*, vol. 3, pp. 404-411.

[20]   L. Uhr, "Layered recognition cone networks that preprocess, classify and describe," *IEEE Transactions on Computers*, vol. 21, pp. 758-768, 1972.

[21]   S. Tanimoto, "A pyramidal approach to parallel processing," in *10th Annual Symposium on Computer Architecture*, Stockholm, Sweden, June 1983.

[22]   D. Hillis, *The connection machine.* Cambridge: MIT Press, 1985.

[23]   J. Rattner, "Concurrent processing: a new direction in scientific computing," *National Computer Conference*, 1985.

[24]   NCube Corp., "Promotional literature," Beaverton, OR, 1985.

[25]   C. Seitz, "The cosmic cube," *Communication of the ACM*, vol. 28, no. 1, pp. 22-33, 1985.

[26]   Sequent Computer System, "Promotional literature," Beaverton, OR, 1986.

[27]   Encore Computer Corp., "Promotional literature," Marlborough, MA, 1986.

[28]   W. Crowther, J. Goodhue, E. Starr, R. Thomas, W. Milliken, and T. Blackadar, "Performance measurements on a 128-node Butterfly parallel processor,," *International Conference on Parallel Processing*, pp. 531-540, 1985.

[29]   G. Pfister, W. Brantley, D. George, S. Harvey, W. Kleinfelder, K. McAuliffe, E. Melton, V. Norton, and J. Weiss, "The IBM research parallel processor prototype (RP3): introduction and architecture," *International Conference on Parallel Processing*, pp. 764-771, 1985.

[30]   D. Kuck, E. Davidson, D. Lawrie, and A. Sameh, "Parallel supercomputing today and the Cedar approach," *Science*, vol. 231, pp. 967-974, 1986.

[31]   H. T. Kung and J. A. Webb, "Global operations on the CMU WARP machine," *Proceedings of 1985 AIAA Computers in Aerospace V Conference*, October 1985.

[32]   T. Gross, H. T. Kung, M. Lam, and J. Webb, "WARP as a machine for low-level vision," in *IEEE International Conference on Robotics and Automation*, ST. Louis, Missouri, pp. 790-800, March 1985.

[33]   H. T. Kung, "Systolic algorithms for the CMU Warp processor," in *Tech. Rep. CMU-CS-84-158, Dept. of Comp. Sci., CMU*, Pittsburgh, PA, September, 1984.

[34]   F. H. Hsu, H. T. Kung, T. Nishizawa, and A. Sussman, "LINC: The link and interconnection chip," in *Tech. Rep., Dept. of Comp. Sci., CMU, CMU-CS-84-159*, Pittsburgh, May 1984.

[35]   M. Annaratone et. al., "The Warp computer : architecture, implementation, and performance," *IEEE transactions on Computers*, December 1987.

[36]   S. Borkar et al., "1WARP: An Integrated Solution to High-Speed Computing," in *Supercomputing Conference*, Orlando, FL, pp. 330-339, November 14-18, 1988.

[37]   F. A. Briggs, K. S. Fu, J. H. Patel, and K. H. Huang, "PM4 - A reconfigurable multiprocessor system for pattern recognition and image processing," *1979 National Computer Conference*, pp. 255-266.

[38]   H. J. Siegel et al., "PASM - a partitionable SIMD/MIMD system for image processing and pattern recognition," *IEEE Transactions on Computers*, vol. C-30, pp. 934-947, December 1981.

[39]   Y. W. Ma and R. Krishnamurti, "The architecture of REPLICA - a special-purpose computer system for active multi-sensory perception of 3_dimensional objects," *Proceedings International Conference on Parallel Processing*, pp. 30-37, 1984.

[40]   W. A. Perkins, "INSPECTOR - A computer vision system that learns to inspect parts," *IEEE Transactions on Pattern Analysis and Machine Intelligence*, vol. PAMI-5 , pp. 584-593, November, 1983.

[41]   C. C. Weems, S. P. Levitan, A. R. Hanson, E. M. Riseman, J. G. Nash, and D. B. Shu, "The image understanding architecture," *COINS Tech. Rep. 87-76,*.

[42]   K. Preston Jr., "Benchmark results: the abingdon cross," in *in Evaluation of Multicomputers for Image Proccessing (Proceedings of the 1984 Multicomputer Workshop, Tuscon, AZ), L. Uhr, K. Preston Jr., S. Levialdi, M. J. B. Duff editors*, Orlando, FL, pp. 23-54, 1986.

[43]   L. Uhr, K. Preston Jr., S. Levialdi, M. J. B. Duff, "Preface in evaluation of multicomputers for image processing," in *Proceedings of the 1984 Multicomputer Workshop, Tuscon, AZ*, Orlando, FL, 1986.

[44]   A. R. Rosenfeld, "A report on the DARPA image understanding architectures workshop," in *Proceedings of the 1987 DARPA Image Understanding Workshop*, Los Angeles, CA, pp. 298-302, 1987.

[45]   M. K. Leung, A. N. Choudhary, J. H. Patel, and T. S. Huang, "Point matching in a time sequence of stereo image pairs and its parallel implementation on a multiprocessor," in *IEEE Workshop on Visual Motion*, Irvine, CA, March 1989.

[46]   M. K. Leung and T. S. Huang, "Point matching in a time sequence of stereo image pairs," in *Tech. Rep., CSL, University of Illinois*, Urbana-Champaign, 1987.

[47] Y. C. Kim and J. K. Aggarwal, "Positioning 3-D objects using stereo images," *Computer and Vision Research Center, The University of Texas at Austin.*

[48] A. Huertas and G. Medioni, "Detection of intensity changes with subpixel accuracy using Laplacian-Gaussian masks," *IEEE Transactions on Pattern Analysis and Machine Intelligence*, vol. PAMI-8, pp. 651-664, September 1986.

[49] F. A. Briggs and E. S. Davidson, "Organization of semiconductor memories for parallel-pipelined processors," *IEEE Transactions on Computers*, pp. 162-169, February 1977.

[50] H. J. Siegel, "Partitioning permutation networks : the underlying theory," *Proceedings of the International Conference on Parallel Processing*, pp. 175-184, 1979.

[51] M. C. Sejnowski et al., "An overview of the Texas reconfigurable array computer," *AFIPS 1980 National Computer Conference*, pp. 631-641, June 1980.

[52] D. Degroot, "Partitioning job structures for SW-banyan networks," *Proceedings of the International Conference on Parallel Processing*, pp. 106-113 , 1979.

[53] E. Horowitz and S. Sahni, *Fundamentals of computer algorithms.* Computer Science Press, 1984.

[54] D. H. Ballard and C. M. Brown, *Computer vision.* Prentice-Hall, 1982.

[55] J. H. Patel, "Analysis of multiprocessors with private cache memories," *IEEE Transactions on Computers*, vol. C-31, pp. 296-304, April 1982.

[56] J. H. Patel, "Performance of processor-memory interconnections for multiprocessors," *IEEE Transactions on Computers*, vol. C-30, pp. 771-780, October 1981.

[57] A. N. Choudhary, S. Das, N. Ahuja, and J. H. Patel, "Surface reconstruction from stereo images : an implementation on a hypercube multiprocessor," in *The Fourth Conference on Hypercubes, Concurrent Computers, and Applications*, Monterey, CA, March 1989.

Index